TECHNOLOGICAL CHANGE AND COMPANY STRATEGIES

ECONOMIC AND SOCIAL ANALYSIS OF TECHNOLOGY

TECHNOLOGICAL CHANGE AND COMPANY STRATEGIES:
Economic and sociological perspectives

Edited by

Rod Coombs
Department of Management Sciences
UMIST
Manchester

Paolo Saviotti
Department of Economics
University of Manchester
Manchester

Vivien Walsh
Department of Management Sciences
UMIST
Manchester

Harcourt Brace Jovanovich, Publishers

London   San Diego   New York   Boston
Sydney   Tokyo   Toronto

ACADEMIC PRESS LIMITED
24–28 Oval Road
London NW1 7DX

United States Edition published by
ACADEMIC PRESS INC.
San Diego, CA 92101

A catalogue record for this book is available from the British Library
ISBN 0–12–187582–2

Typeset by J&L Composition Ltd, Filey, North Yorkshire
and printed in Great Britain by Hartnolls Ltd, Bodmin, Cornwall

CONTENTS

LIST OF CONTRIBUTORS

MARK BODEN Programme of Policy Research in Engineering, Science and Technology, University of Manchester, Manchester M13 9PL, UK.

MICHEL CALLON Centre de Sociologie de l'Innovation, Ecole Normale Supérieure des Mines, 62 Boulevard St Michel, 75272 Paris, Cedex 06 France.

ROD COOMBS Manchester School of Management, University of Manchester Institute of Science and Technology, PO Box 88, Manchester M60 1QD, UK.

ISTEMI DEMIRAG Sheffield University Management School, PO Box 598, Crookesmoor Building, Conduit Road, Sheffield S10 1FL, UK.

MARK DODGSON Science Policy Research Unit, University of Sussex, Brighton BN1 9RF, UK.

KENNETH GREEN Manchester School of Management, University of Manchester Institute of Science and Technology, PO Box 88, Manchester M60 1QD, UK.

PETER GROENEWEGEN Vakgroep Algemene Vorming, Vrije Universiteit, De Boelelaan 1083, 1081 HV Amsterdam, The Netherlands.

RAY LOVERIDGE Strategic Management and Policy Studies Division, Aston Business School, Aston University, Birmingham B4 7ET, UK.

DONALD MACKENZIE Department of Sociology, University of Edinburgh, 18, Buccleuch Place, Edinburgh EH8 9LN, UK.

STAN METCALFE Department of Economics, University of Manchester, Manchester M13 9PL, UK.

KAREL MULDER University of Groningen, Management and Assessment of Technology, Nijenborgh 16, 9747 AG Groningen, The Netherlands.

PAOLO SAVIOTTI Department of Economics, University of Manchester, Manchester M13 9PL, UK.

JOHAN SCHOT Centre for Technology and Policy Studies, Netherlands Organisation for Applied Scientific Research (TNO), PO Box 541, 7300 AM Apeldoorn, The Netherlands.

ANDREW TYLECOTE Sheffield University Management School, PO Box 598, Crookesmoor Building, Conduit Road, Sheffield S10 1FL, UK.

PHILIP VERGRAGT Ministry of the Environment, PO Box 450, 2260 MB Leidschendam, The Netherlands.

VIVIEN WALSH Manchester School of Management, University of Manchester Institute of Science and Technology, PO Box 88, Manchester M60 1QD, UK.

PREFACE

This volume is the first in a series which sets out to serve as a vehicle for research and scholarship on science and technology, operating explicitly at the interface between social science disciplines and reflecting our conviction that the study of the roles of science and technology in modern industrial societies has now, even more than ever before, reached a point where substantial progress is made principally through such multidisciplinary and interdisciplinary approaches.

The editors of this volume collaborated in the mid-1980s as authors of a book on economics and technological change, in which we took faltering steps toward integrating some aspects of organization and management writing within the basically neo-Schumpeterian and evolutionary perspective of that book. It is a tribute to the work of our colleagues in the field of sociology of technology, and to the open-mindedness of those working in a more economic tradition, that the passage of only six years has seen such major steps forward in the cross-fertilization of sociological and economic perspectives on technology.

In this first volume, the aspect of science and technology which is under scrutiny is its strategic deployment by business firms. The relevant disciplines which are brought into engagement with this topic, and with each other, are economics and sociology. Future volumes in the series will address different topics and involve other disciplines.

The contributions to the book have been selected from a larger collection of papers which were presented at an international seminar which was organized by the editors in Manchester in September 1990. The contributors have taken account of the discussion at that conference in revising their papers for this volume, and we are extremely grateful both to them for the careful attention they have given to this task, and to the discussants at the conference who provided detailed comments to the contributors. We would also like to record our gratitude to our colleague Albert Richards who played a very major part in the organization of the Manchester seminar.

Rod Coombs, Paolo Saviotti, and Vivien Walsh

1

TECHNOLOGY AND THE FIRM: THE CONVERGENCE OF ECONOMIC AND SOCIOLOGICAL APPROACHES?

Rod Coombs, Paolo Saviotti and Vivien Walsh

INTRODUCTION

The theme of this book and of the conference from which it arose is that the explanation of two related sets of phenomena, firm strategies and technological change, cannot be wholly contained within one discipline, but must cross the borders between disciplines. The relevant disciplines (or subsets of disciplines) considered here are microeconomics and microsociology. This introductory chapter attempts to compare and contrast the explanations of firm strategies and technological change given by each of the two disciplines. Some preliminary observations on interdisciplinary differences will be used to structure the comparison. These will be followed by a brief analysis of the historical development of the approach of the disciplines to the phenomena of firm strategy and technical change. Next we address the questions of whether, in the limited context of the analysis of these phenomena, the two disciplines are convergent or not, and of what outcomes can be expected from the present trends.

Finally, having set the volume in this general context, we give an overview of the other contributions, some of which are theoretical and others derived from case studies, discussing their content, conclusions and relevance for the present debate.

Disciplines and research traditions differ in terms of both the object of study and the tools that they use to study it. Different objects of study define different disciplines. In general, however, this is not the only difference between disciplines. Each object of study presents a peculiar set of problems and leads to the development of different concepts, techniques and approaches. This explanation of interdisciplinary differences is essentially functional: different objects of study "cause" different tools and approaches. But naturally the use of different concepts, techniques and approaches can also be the result of other factors, such as the social relations within and between disciplines. This can be further complicated if the objects of study of two disciplines have a substantial overlap. This type of analysis cannot be developed in detail here. However, for the purposes of the present book it alerts us to the fact that we cannot regard any particular definition of the objects of study or tools of analysis of economics and sociology as "natural" or fixed.

Even within a given discipline the object of study does not remain constant with time. New phenomena may appear or be incorporated into a discipline, new modes of explanation may merge and replace pre-existing ones. Let us then briefly examine the development of the approaches adopted by microeconomics and by microsociology to the phenomena of technical change and of firm strategy.

TECHNICAL CHANGE IN ECONOMICS AND IN SOCIOLOGY

Changes in the way in which technical change is conceptualized have occurred in both disciplines. A brief outline of these changes will be given here. We begin with a stereotyped history of the conceptualization of technical change in economics and in sociology, contrasting it subsequently with modern developments in both cases.

First of all it must be observed that the assumption of *homo economicus*, as separate from *homo eticus*, *homo religiosus* and all other *homines* (Georgescu Roegen, 1971, p. 318) establishes very rigid boundaries between economics and the other social sciences. This division is not limited to the analysis of technical change, but is common to the whole of neoclassical economics.

In the dominant research tradition in economics (often referred to as neoclassical, orthodox or marginalist economics), technical change was traditionally represented as exogenous to the economic system. Furthermore, it is only the effects of technical change on economic variables, such as productivity, prices, etc., which are analysed by the theory. The process of generation of technical change and its intrinsic features are excluded from the representation. These are the basic features of what could be called the "production function approach" (Coombs et al., 1987). In the course of time one expects the isoquants of the existing production function to move towards the origin, showing a greater efficiency of existing processes. Neither the internal features of these processes, nor the nature of the output produced, are included in this representation.

Understandably this approach has a number of limitations, mainly in terms of its ability to give indications about policies for science and technology. For example, it would be very difficult to decide what product lines to develop, and therefore what R and D projects to sponsor, purely on the basis of this theoretical structure. According to the neoclassical approach, the firm was considered as a profit maximizer, as an organization having perfect information, and one in which owner and entrepreneur were the same person. The firm, like technical change, was considered a black box and its role was essentially to provide conceptual support for the existence of markets (Coombs *et al.*, 1987; Moss, 1981). The presence and effect of internal structures and the concept of strategy could not be accommodated within this stylized and highly oversimplified representation.

The limitations of this approach have been evident for a number of years and have given rise to new developments in the analysis of technical change. These alternative approaches have generally remained minority points of view, but have been steadily gaining ground.

To begin with, a number of detailed studies of the process of technological innovation have been conducted from the 1960s onwards, which have, usually in a very empirical way, approached such problems as: what are the factors leading to successful innovations? Such questions could not be asked within the dominant research tradition in economics. These studies greatly enriched our picture of technical change, but ran into the predictable obstacles of non-comparability and of non-generalizability of their results (Coombs *et al.*, 1987). The outcome of these innovation studies has been to give rise to a number of attempts to create a theoretical structure which encompasses their results. Many of the contributors to innovation studies are now amongst the proponents of an evolutionary theory of economic and technological change (Nelson and Winter, 1982; Dosi *et al.*, 1988).

Modern evolutionary theories of economic and technological change bring together a number of research traditions, including a dissenting one within economics (represented chiefly, but not only, by Schumpeter), organization theories, biology, systems theory, and irreversible thermodynamics (Saviotti and Metcalfe, 1991).

Schumpeter had clearly foreseen the dynamic, non-equilibrium character of economic development, and the fundamental role played in it by innovations (Schumpeter, 1934, 1943). However, both as a result of his own personal background and due to the time in which he was living, he did not incorporate a number of institutional features which characterize modern capitalism (Nelson, 1990). However, if a paternity of evolutionary theories is to be attributed to anyone, it is undoubtedly Schumpeter.

For the present discussion, we can include within organization theory non-neoclassical theories of the firm, research in business history, and studies of a more managerial type, centred around the concept of strategy. Simon's concept of satisficing rather than optimizing behaviour, his distinction between substantial and procedural rationality (1969, 1981), and Cyert and March's (1963) behavioural theory of the firm, with the important role played within it by groups with conflicting interests and goals, introduce uncertainty into firms' operations, and open up the black box of the firm. The concepts of strategy and structure play a fundamental role in the research tradition of business history, which draws its inspiration from the work of Chandler (1962, 1977). Williamson's transaction cost analysis (1975, 1985) provides a more analytical framework, addressing similar themes to many of the concepts described above.

Following this emerging research tradition, authors like Kay (1982) and Teece (1986, 1990) have addressed issues of firm strategy and of technological innovation. In particular, the concept of corporate capabilities in Teece's work (1990), bears a close resemblance to that of knowledge base used by other authors (Metcalfe and Gibbons, 1989).

In summary it could be said that developments in the analysis of technical change and of the firm have moved in the direction of a greater realism, for the moment at the expense of analytical accuracy. The black boxes of technology (Rosenberg, 1982) and of the firm have been opened up. The traditional point of view, in its formal elegance, has been surrounded by a number of studies and theories which, in their less accomplished formalism, but with their greater relevance, challenge the core of the theory.

The present phase could thus be described as one of knowledge in transition. Alternatively, referring back to the discussion in the introduction to this chapter, it could be said that the object of study of economics, or rather that part of the object of study which concerns both technical change and firm theory, has been extended by the inclusion of new phenomena. These new phenomena include the internal structure of the firm, the acquired knowledge and ability of the firm, the rules of behaviour used within the firm, and the processes by which the firm interprets its options and chooses between them. Interpretation of these new phenomena may require new tools.

In sociology, attention has been paid to the role of technology since at least the 1950s. A first major feature of the treatment of technology in the sociology of organizations was the relationship between technology type and organization structure. The pioneering work of Joan Woodward (1965) established the presence of a correlation between the type of technology used (for example mass, batch or flow) and the organizational structure adopted. The conclusions reached here, that certain technologies, if they are "core" technologies for the business or organization, predispose an organization to favour certain structures and eschew others, is obviously broadly in line with some form of contingency theory. Another pole in the argument is the "strategic choice" paradigm associated with Child (1972), which retains a contingency flavour but uses contingencies to set limits on action, rather than to determine it. These early attempts, in a way which paralleled what was happening in economics, did not deal with the provenance of technologies, but only with their supposed effects.

A second feature of the sociological discussions of technology is the analysis of technology in relation to the organization and control of work. Again there have been two rather opposed traditions here. First there has been the socio-technical systems approach (Emery and Trist, 1960, 1965), in which the "implementation" of new technology and its effect on work redesign and control were studied. The issues were seen here as task design and allocation, in order to optimize both the efficient use of technology, and worker satisfaction. The most characteristic feature of the socio-technical systems school is the exclusion of political economy considerations and the assumption of the possibility of rational consensus. With respect to technology, the focus was still mostly on effects rather than provenance, though the more radical variants did address employee involvement in system design.

The other research tradition here is the labour process tradition (Braverman, 1974). This began from a Marxian perspective which foregrounded political economy considerations and denied consensus. The implication of this approach for the provenance of technology was that objectives of exploitation, variously interpreted as valorization or direct control of workers, were seen as ever-present influences on the process which generates technology. Effects were seen to follow naturally from these inbuilt biases.

Almost in parallel with these research traditions in the 1960s began the development of a sociology of science (see for example Barnes, 1972) which highlighted the procedures and determinants of scientists' behaviour, and the limitations of the so-called scientific method. This tradition's predominant concentration on science in the 1960s and 1970s has given way to a much greater emphasis on the sociology of technology in the 1980s (Callon et al., 1986; Bijker et al., 1986). Increasingly these developments have led scholars to pay greater attention to the internal structure of technologies (see, for example, Mangematin and Callon, 1991).

In many respects the perspectives outlined in sociology thus far can be seen to mirror, albeit from another vantage point, the exogenous approach of early economic theories. The technology is given, and the object of study is its effects. Only since the latter part of the labour process work has the plasticity of technology come more into focus.

RECENT DEVELOPMENTS

So far, this chapter has provided an overview of the development of economic and sociological approaches to technical change and to the firm since the 1950s. As has been pointed out, in both cases there has been a trend towards a growing realism and a more explicit analysis of both technology and the firm. These trends have been parallel in the two disciplines, and have led to a greater similarity of their object of study. The problem thus arises of whether the two disciplines, in the domain of the analysis of technical change and of the firm, are converging. In order to put this question in sharper focus we need to examine some recent developments on both sides.

Evolutionary economics starts by combining the Schumpeterian heritage, stressing the fundamental role of innovations in long-term economic development (Schumpeter, 1934), with the behavioural theory of the firm (Cyert and March, 1963). Economic and technological changes are, therefore, introduced and implemented by satisficing, rather than optimizing, firms (Nelson and Winter, 1982). Two mechanisms, variation and selection, are at the basis of economic evolution. Variation, the generation of new species, is the outcome of firms' search activities. Selection occurs by means of competition, though this is seen as taking place in a "selection environment", rather than a simple neo-classical type of market.

Firms' behaviour is then determined by the balance between search activities and the operation of the selection environment. However, search activities do not scan the whole environment, but are generally guided by previous experience. Firms' knowledge is therefore constrained and local. Recent studies have shown that, both within and outside the firm, patterns and rigidities exist which constrain search activities and overall firm behaviour. Routines and decision rules within the firm (Nelson and Winter, 1982), technological trajectories and regimes (Nelson and Winter, 1977), dominant designs (Abernathy and

Utterback, 1975, 1978), technological guideposts (Sahal, 1981a, 1981b), and technological paradigms (Dosi, 1982) at a higher level of aggregation provide examples of attempts to conceptualize these patterns.

The nature of knowledge in organizations has also become an important component of modern evolutionary theories. Thus knowledge can be characterized as tacit or codified (Polanyi, 1962, Teece, 1981; Nelson and Winter, 1982), public or private, local (Nelson and Winter, 1982) and cumulative (Pavitt, 1990). An important outcome of these considerations is that a firm's knowledge is highly (although not entirely) specific and path dependent (see Metcalfe and Boden, this volume). From a different point of view, irreversibility and path dependency can be seen as the consequences of increasing returns to adoption of technologies (Arthur, 1983, 1988, 1989).

So much for recent developments in evolutionary economics. In sociology, an interesting and important recent development has been constituted by the actor-network theory (Callon *et al.*, 1986). The networks in this theory are different from those usually encountered, in that they are constituted by humans, by non-humans (e.g. scientific equipment), by inscriptions, and by money. Thus they are heterogeneous "techno-economic networks" (TENs). TENs are constituted by actors and intermediaries. Actors "translate" other actors and intermediaries. The translation is an interpretive operation in which an Actor A defines other actors/intermediaries, B, C, D, . . . In turn B, C, D, . . . are interdefined, that is linked together, by the translation. In terms of older sociological language one can therefore think of TENs as the shifting framework within which a particular aspect of reality is "socially constructed"; but with the added complication of non-human actors. (For a detailed account see Callon, this volume.)

The networks thus formed can differ both in their convergence and in their degree of irreversibility. In highly convergent and irreversible networks the actors are perfectly identifiable and their behaviour is known and predictable. Normalization is another important feature of TENs. In a normalized TEN, links are more predictable, fluctuations more limited, actors and intermediaries more closely aligned, and less information is put into circulation. Thus the possibility of analytically describing TENs depends on their degree of irreversibility, of convergence and of normalization. TENs are not rigid and fixed forever. They can break down and give rise to different TENs. For sociologists using this theory, the analysis of concrete instances of technological change is conducted through the analysis of the associated networks and their changing properties of reversibility and normalization.

The possible existence of convergence of economic and sociological accounts of technical change can now be briefly analysed on the basis of these summaries of recent developments. First, as already mentioned, there has been a trend towards a greater degree of realism and a more explicit representation of technology and of the firm. Within a broadly economics tradition, studies of diffusion (Metcalfe, 1981, 1988), of competition (Metcalfe and Gibbons, 1989) and of technological evolution (Saviotti and Metcalfe, 1984; Saviotti, 1988, 1991) have incorporated product quality or product characteristics within analytical models. Therefore, the object of analysis of the two disciplines has acquired a greater similarity. Second, some of the tools used in the analysis have become more similar. The concepts, from routines to paradigms, which have emerged in evolutionary economics and which describe patterns and stabilities in firms and technologies, have already been described. In sociology a parallel to routines is provided by the concept of cultural styles of decision making (Farmer and Matthews, 1990). Such a concept is not new, but it can find both a reinforcement and an easier bridge to economic analysis in the actor-network theory mentioned before. Specifically, the actor-network theory acts as a counterweight to any technological determinist interpretation of such concepts as "technological trajectories" mentioned above. By constitut-ing the trends in the evolution of technologies as social institutions rather than "natural laws", it permits the routines and paradigms within firms to change by means other than a life-or-death survival process achieved through a selection environment.

There is then a degree of commonality among these conceptual approaches, in that they all imply a degree of fine structure in technological and organizational features of the world under study. This structure has a short- to medium-term stability, but it can eventually change by means of transitions to other states and structures. Put at its starkest, we may suggest that where the evolutionary economist sees a stable natural trajectory, the sociologist sees a normalized irreversible techno-economic network: where the evolutionary economist sees a radical innovation, the sociologist sees ruptured networks and the emergence of new networks.

FIRM STRATEGY, TECHNOLOGICAL CHANGE AND CORPORATE CAPABILITIES

The previous sections have dealt with the changing perceptions and approaches to technological change and to firm theory in economics and in sociology. These are closely connected to the problem of firm strategy, and it is to this that we now turn.

We have first to observe that there is no place for strategy in the orthodox theory of the firm. In the presence of perfect knowledge, firms possess algorithms which allow them to calculate optimum solutions. There is no uncertainty and no choice — or, at least, choice is purely mechanical (Shackle, 1955; Hodgson, 1991). The need for strategy arises in the presence of uncertainty, of imperfectly perceived and understood multiple outcomes, and of several ways by which to achieve each outcome.

In a different way, the possibility of strategy can be explored by reference to the dichotomy between the internal and the external (or selection) environment of the firm. In the presence of a given external environment, the firm may be thought of as attempting to adapt to it. Furthermore, firms often tend to modify the environment to improve their own chances of survival. This has two consequences: first, strategy can be defined as the selection and implementation of a set of goals aimed at adapting or at modifying the external environment; second, differential adaptation or differential modification lead to degrees of performance and success. In this context the relationship between strategy, structure and tactics is related to that between the internal and the external environment of the firm. The internal environment is itself sufficiently complex to give rise to a hierarchical structure (Simon, 1981), and, therefore, to achieve self-regulation. Adaptation to the external environment takes place without a complete change of internal structure, or at least this happens in a large proportion of circumstances. The invariance of some parts of the internal environment with respect to external changes is an example of homeostasis, a common property of complex systems. In a recent analysis which has strong resonances with this argument, Pavitt (1990) has shown that large R and D-intensive firms display extraordinary institutional continuity in the face of technological discontinuities.

This discussion of the internal and of the external environment allows us to proceed to analyse different conceptions of strategy. According to Teece et al. (1990) there are four different concepts of strategy which can be identified. First, there is the concept developed by Porter, which can be considered a variant of the structure-conduct-performance approach. According to Porter (1980) the essence of strategy consists in relating a company to its environment. The relevant part of the environment is constituted by the industry of which the firm is a member. This environment can be characterized by five basic competitive forces: entry, threat of substitution, bargaining power of buyers, bargaining power of suppliers, and rivalry among current competitors. Of these forces, four are related to the external environment of the firm and only one, the rivalry among current competitors, involves the analysis of firms' capabilities. Competitor analysis can be performed by comparing each competitor's strengths and weaknesses with respect to a number of functions, which are in turn related to the five key competitive forces. While in Porter's analysis there is some recognition of firm-specific assets, according to Teece differences among firms are highly stylized, and relate primarily to scale.

The second concept of strategy is that of Shapiro (1989), and it is based both on the new industrial organization and on game theory. In this context a strategic move, such as a commitment, threat or promise, is designed to influence the behaviour of others. A strategic move is successful if it influences the behaviour of others, either by influencing competitors' pay-offs or simply by influencing their beliefs. The main analytic tool used to explore these issues is game theory. From the viewpoint of this book, the game theoretic entry deterrence approach has two relevant features: first, it has no theory of the firm, and second, it sees competitive advantage stemming from deceit and from restrictive practices. The emphasis of this approach is on a short-run, compete-through-existing assets view of strategy. In summary, in both Porter's and the strategic entry deterrence approaches, the emphasis is on the external rather than the internal environment of the firm.

The third perspective on strategy is the resource-based one. This approach focuses on the rents accruing to the owners of scarce firm-specific resources, rather than on the economic profits from product market positioning. This research tradition can be related to the work of Penrose (1959) and of Williamson (1975), and to their emphasis on capturing rents on scarce, firm-specific assets, whose services are difficult to sell in intermediate markets. In this perspective a successful strategy consists of exploiting the specific resources/capabilities/endowments of each firm. Such capabilities are not only heterogeneous and firm specific, but also very difficult to create. Teece et al. (1990) propose the dynamic capabilities approach as a variant of the resource-based approach, placing greater emphasis on the creation of new capabilities, as opposed to the exploitation of existing ones.

In the dynamic capabilities approach, the development of new capabilities occurs through firms' learning in the presence of a number of constraints. Learning processes are essentially social and collective. They generate knowledge which resides in organizational routines (Nelson and Winter, 1982). Path dependency, complementary assets, and transaction costs constrain learning processes. Learning is local and the past experience of a firm is likely to condition its present and immediate future development. Complementary assets enable a firm to develop certain types of capability. The more firm-specific assets are, the more likely it is that a firm has to invest internally to generate them.

To summarize, while the industry structure and the entry deterrence approaches focus more on the constraining effects of the external environment, the resource and dynamic capabilities approaches focus predominantly on the internal environment of the firm. In this sense these approaches may be complementary, rather than exclusive.

The most important point for the purposes of the present discussion is the role of knowledge in the generation of capabilities. First, new knowledge is generated through a collective process, to then constitute a firm's knowledge base (Metcalfe and Gibbons, 1989). Such a knowledge base is partly tacit and partly codified, and is difficult to imitate in its entirety. It is the collective nature of this process which makes it suitable for analysis, from both an economic and a sociological point of view. On the one hand, knowledge is generated through a trial and error procedure, but always remains imperfect and firm specific. On the other hand, it is related to present and to past power structures, and to patterns of interaction within the firm, and across firm boundaries.

Turning again to sociological and oganization-theoretical construc-tions of the topic of strategy, it is possible to discern some parallel features in the evolution of the literature. The early normative accounts of strategy which promised firms' rational planning of their internal and external environments were quickly replaced by more "realistic" and descriptive accounts. Thus Mintzberg (1978) and Johnson and Scholes (1984) emphasize the incremental and iterative nature of strategy-making in the face of uncertainty. Pettigrew proposes (1987) a "processual" model in which there is a continual interaction between the "content, context and process" of strategy. Subsequently, critical accounts of strategy (Knights and Morgan, 1990) have drawn attention to the sense in which the very act of framing a strategic analysis and voicing it can itself empower the group concerned within the organization. This analysis has resonances with Callon's notion of "translation" (this volume), inasmuch as the strategy can be seen as an intermediary which translates other actors in the organization, and indeed translates the organization itself.

Thus we can glimpse emergent connections between the power dimensions of "making" strategy; the collective and path-dependent process of acquiring knowledge and dynamic capabilities specific to a firm; and the multiple actors, claims, and moves involved in normalizing a techno-economic network.

IS THERE A CONVERGENCE BETWEEN ECONOMIC AND SOCIOLOGICAL APPROACHES?

As we have seen in the analyses of technological change and of firm theory and strategy, economics and sociology have proceeded along parallel lines. In both cases, recent trends have led to a more explicit analysis of the phenomena under study, which could be called an opening of black boxes. This is not entirely accurate, because sociology has always had an interest in the internal realities of firms, but it is generally a suitable description of the common trends followed. There has been a convergence of the objects of analysis of both disciplines, at least with respect to firm strategy and technical change. In both cases the internal structures of firm strategy and of technical change have become part of the object of analysis of the disciplines. In some senses this can be represented as a temporary convergence. However, disciplines are not characterized only by their object of analysis, but differ in a series of other aspects, such as the specific tools that they use. Even if there is a temporary convergence of their objects of analysis, it is possible for them to diverge in the longer term, if their modes of analysis develop in incompatible ways. At present, however, there is some substantive overlap between the tools and objects of evolutionary theories of firms and technologies; of managerial theories of the strategic attributes of firms; and of actor-network accounts of the social and institutional provenance of technologies. The contributions to this volume all exemplify this area of overlap in various ways. In the remainder of this introductory chapter we briefly summarize the contributions, and situate them with respect to the agenda sketched above.

SYNOPSIS OF THE BOOK

One of the main themes of this book, the suitability and compatibility of microeconomics and microsociology for the analysis of technical change, is approached directly in chapter 2 by Donald MacKenzie. He draws parallels between the major focus of evolutionary economists on the uncertainty inherent in technological change, and the interpretive flexibility (existing particularly at times of scientific controversy) which is a central theme of the sociology of scientific knowledge. Similarly, he draws parallels between successful innovation in technology, and "closure" in science — the victory of one of several competing scientific theories, which then becomes "established science". The processes which lead to the emergence of the apparently self-sustaining realms of "objective" scientific knowledge on the one hand, and economic processes on the other, offer parallel topics for analysis and mutual enlightenment.

MacKenzie suggests that while an understanding of the causes of persistent patterns of technological change offers the potential for bridge-building between economic and sociological analyses of technological change, he is critical of the now widely-used metaphor of the technological trajectory (and especially the "natural" trajectory) in the alternative or evolutionary economics literature. The trajectory concept reflects a justifiable reaction against the idea that technology is an entirely plastic entity, shaped at will by market forces or other social factors, but he believes it moves too much in the other direction of technological determinism and "internalism".

MacKenzie's reservations about trajectories are that technological development is essentially a social process, and should not be described as "natural". Technology does not develop a momentum "of its own"; there is an element of self-fulfilling prophecy involved. Persistent patterns of technological change are persistent partly because technologists and others believe they will be persistent, and it is through the process of adoption that technologies become irreversibly superior. Thus MacKenzie argues that a technological trajectory is an institution, in the sociological sense of being sustained not through its internal logic or intrinsic superiority, but because interests develop in its survival and development, which reinforce and are reinforced by the belief that it will continue; and that it could be more usefully analysed by the actor-network approach of Callon, Latour and others, as indeed Callon does in chapter 4.

Another potential bridge described by MacKenzie is "ethnoaccountancy", the study of the way in which people in different cultures and historical periods do their financial reckoning (in this case of technological change). Accounting is clearly related to economic analysis since firms use measures of costs, incomes, profits and so on in making decisions about investment in technological projects and assessing their relative performance. It is also clearly related to sociological analysis too, in that accounting practices and definitions of profit are not established once and for all. Present-day practices and definitions evolved as an inseparable part of the emergence of the modern business enterprise, and there is some evidence to suggest quite different emphases in Japan from those in the USA and UK. Different techniques of financial assessment result in apparently objective, numerical support for either short-term or long-term strategies towards technological change, or more labour intensive or capital intensive investment strategies. The study of these activities provides a potentially very promising way of linking sociological and economic approaches; Tylecote and Demirag pursue short termism further, in chapter 9.

The concept of strategy formation plays a central role in chapter 3, by Stan Metcalfe and Mark Boden. Their contribution is a development and elaboration of evolutionary economics, making use of concepts often to be found in the realm of sociology, such as strategy formation (though sociologists may prefer notions such as the "seamless web" — indicating that the content of technologies are shaped simultaneously with the context — to the separation here of internal and external environments). Strategy formation is presented as a dynamic process whereby a firm explores its technological and market environments, and acquires, organizes and uses knowledge in gaining competitive advantage. Their theme is the emergence of strategy at the intersection of three categories of phenomena: competitive behaviour, organizational design, and technology content and structure. The firm must balance between the internal selection environment it generates and which is embodied in its decision rules and communication structures, and the external selection environment it faces. Its perception of the external environment, meanwhile, is also conditioned by the internal environment (a point taken up later by Green and by Vergragt *et al.*). The firm must also strike a balance between its current range of activities and the generation of activities necessary to maintain its position in the future (chosen from among the options determined by the internal selection environment).

Central issues in evolutionary change are the sources of variety on which selection mechanisms operate, and the sources of limits to the scale and scope of variety. The authors elaborate the ways in which both are connected to the formulation of strategy, and the process by which it is implemented in the firm. They emphasize that biological evolution is only a metaphor, since economic evolution, though similarly driven by variety, is a faster process. More importantly, economic evolution allows for imitation and has a "Lamarckian" element of firms learning and seeking to modify the selection environment in their favour. (The learning process is further developed in Mark Dodgson's chapter.)

Metcalfe and Boden use the concepts of trajectory and paradigm, defining paradigm in such a way that some of MacKenzie's objections are accommodated; that is, as an accumulation of technological capability which is not random, but structured by technological and non-technological factors. They find the concept of paradigm useful in their analysis, because it "permits variety in outcomes, while placing limits on permissible variety". Their criticism of the paradigm concept is that it allows these limits only to be transcended by a "revolution". They argue that a radically improved technology may replace mature rivals not by sudden changes but by a gradual accumulation of new knowledge and equipment, and a growing belief in and demonstration of its superiority and potential for solving problems.

Metcalfe and Boden's other criticism of the paradigm as used, for example, by Dosi, is that its organizational context is not specific. (The importance of the firm is a point also developed later by Green.) They develop a new concept, that of the "strategic paradigm", which connects business opportunities and technological capabilities, and takes on board the context of the specific business unit. Differences in the long-term competitive performance of firms are then shown to be inextricably linked with these strategic paradigms. Although evolutionary change is driven by technological variety, it is the creative strategies of firms which stimulate that process. The paradigm generates variety, but it also constrains the kinds of variety which can be considered. Metcalfe and Boden also use the idea of "design configurations", which have some of the qualities of a paradigm but are more narrowly focused. Business units specialize around individual design configurations, which are the fundamental units around which the authors organize discussion of technological change.

Michel Callon, in chapter 4, is not averse to the concept of technological trajectory, but prefers to write in terms of irreversibility and techno-economic networks. Irreversibility is the process where certain trajectories stabilize and are successful, instead of seeing a variety of new configurations. His analytical framework allows the charting of an evolutionary process in which a network passes from a state of flux and divergence to one of strong irreversibilization and standardization. This techno-economic network is a co-ordinated set of heterogeneous actors, including public laboratories, centres for technical research, companies, financial institutions, users and government bodies. These actors participate collectively in the conception, development, production, distribution and diffusion of products, processes and services, some of which give rise to market transactions. Callon's aim is to analyse the interactions between these actors in such a way as to account for the choices made in creating science and technology and in diffusing and consolidating its results.

The techno-economic network has three poles: those of the market, science and technology, with many links between them. They are not just pure associations of human beings. Intermediaries are whatever passes between them, for example texts of various kinds (papers, patents, blueprints, research proposals), artefacts (components, software, equipment), people with skills, and money in various forms. Economists are familiar with the idea of things bringing actors together: for example, the consumer and producer is linked via the product. Sociologists work on the basis of actors being inseparable from their social relations. Callon's concept of the techno-economic network draws both of these approaches together, by showing how the interaction of actors materializes in the intermediaries that circulate between them.

He also shows that any of the intermediaries can be seen as networks with social and technical components. Thus a text defines items like enzymes or electrons and scientific procedures, and also funding agencies and companies, and the associations between them all. Technical artefacts are linked to users, repairers, diffusers, the skills needed to use them, their performance characteristics, and how and by whom maintenance will be done. Skills are associated in networks with equipment and with training, recruitment and operating procedures. Money may represent information or commitment, and is linked to politicians setting exchange rates, banks and funding bodies. Intermediaries may also be hybrids of any of these.

Callon uses the concept of "translation" essentially to mean the assertion of a definition of something, with possible real effects on the thing defined; for example, a scientific text translates a monoclonal antibody by setting down its attributes. At the same time this translation might also mean issuing a polemic against alternative translations, or even establishing that a particular research strategy and certain industrial developments are necessary. Translation can also suggest that something is presented in a language that is meaningful to the intended reader (for example, technological opportunities are translated into company strategy). Elsewhere Callon (1986) discusses and defines four "moments" of translation: problematization, *interessement*, enrolment and mobilization. In Callon's account, networks possess differing degrees of "convergence" — the degree of accord resulting from a series of translations, which can increase or decrease. Thus the degree of "irreversibilization" means the degree to which it becomes impossible, as a result of translation, to return to a situation where the next step is only one possible option among others. It also means the extent to which later translations are pre-determined. It is a result of the durability and robustness of the intermediaries, and depends on the strength of association in the networks.

The concept of learning is central (see also Dodgson's chapter). Through progressive mutual adaptation, the different elements involved in a translation become exclusively dependent on each other. Decisions become more and more dependent on the history of past translations. Normalization of behaviour then renders certain links predictable, limits fluctuations, aligns actors and intermediaries, and cuts down on the number of translations and the amount of information put into circulation. Unlike trajectories, networks can rarely be separated into simple and easily quantifiable descriptive frameworks. Networks which are not very convergent or irreversibilized correspond to conditions of uncertainty.

Callon's treatment of non-humans (like electrons, enzymes, money, patents and, elsewhere, scallops) in a similar way to humans (not "privileging the social") is not intended to be anthropomorphic — the fact that they intervene in processes does not mean they have motives and intentions of their own. It is intended to underline, as MacKenzie has also done, the fact that technology is not infinitely plastic, to be shaped in any way whatsoever by social forces, any more than technology is driven solely by its own internal logic, independently of society.

Ray Loveridge in chapter 5 explores the trajectory concept in a different way again, focusing on life cycles of technologies and industries, as well as learning curves and diffusion patterns, all of which are commonly described in the literature as following an S-shaped curve. In particular, he is interested in periodic discontinuities resulting in crises of control within individual firms and sectors, and he provides a critique of a very wide range of literature in which crisis metaphors are used. Crises, he argues, are often precipitated by changes in perception and expectations, rather than necessarily by a shift in "objective reality", a point touched upon by several authors in this book. Loveridge speaks of "architectural change" (in the organization and governance of firms and their networks of relationships) involving for example the disappearance of old customers, suppliers and competitors, and the emergence of new, quite different forms. At points of such architectural changes, he observes that the way in which human assets (especially "scarce and idiosyncratic competences") are recognized, developed and institutionalized, are crucial to the future viability of the new corporate hierarchies and structures.

Loveridge uses the notion of "cognitive paradigm" in a similar way to Metcalfe and Boden's "strategic paradigm"; not just referring to science or technology, but also to a firm's modes of achieving its objectives by gaining and exercising control over internal and external relationships. His conceptual approach is illustrated by comparing the firms Lucas and Bosch over an historical period of adaptation to post-Fordist organization. He observes a strongly resilient pattern of strategic behaviour towards technological development and exploitation, with roots in the founding of the firms and their early adoption of particular corporate structures many years before.

Both Mark Dodgson (chapter 6) and Kenneth Green (chapter 7) have chosen as a case study the development and commercialization of biotechnology, although they use it to explore different aspects of firm strategy. Dodgson discusses the case of one firm, Celltech, and his central theme is learning as an active strategy for commercial exploitation of a novel technology. As a new small firm Celltech evolved its own appropriate management systems and practices, and new organizational arrangements. It developed an adaptive business strategy which was able to amend its aims in line with changing technological and competitive circumstances, and its own developing competences. Technological and organizational learning is a topic analysed in the innovation studies, industrial economics, organizational behaviour and management literatures. Dodgson draws on behavioural theories of the firm, such as Cyert and March's work, the institutional changes which enable the take-off of new techno-economic paradigms according to the perspective of Freeman and Perez, and Rosenberg's learning-by-doing and learning-by-using, among other approaches. He develops a multi-faceted analysis of his central theme of learning as a strategic and tactical activity, whereby the firm generates and acquires technological and market knowledge and modifies its own behaviour accordingly.

In chapter 7 Kenneth Green shows that firms developing and commercializing products from a radically new technology are unable to respond to market signals, since no measurable kind of demand in the economic sense exists. There is no question, either practically or conceptually, of constructing an innovative product which diffuses into a pre-existing external socio-economic environment, on which in due course it has some kind of impact. Instead, such firms are actively engaged in shaping the market, creating a "market space" as he calls it, as well as shaping the technology. (Or in some cases, failing to create a market space.) In the case of the monoclonal antibody-based diagnostics and genetically engineered hormones which Green describes, innovating firms had to (or had failed to) enlist the support of regulatory agencies, the medical profession, farmers and environmental groups.

The continuous acts of moulding the environment so as to facilitate the innovation then, in turn, modify the technology-shaping activities. Green draws on economic ideas, such as Metcalfe's development of diffusion theory, in which the profitability of the supplier and the demands of the customer mutually interact, and Walsh's changing patterns of "demand-pull" and "discovery-push" over the life cycle of a technology or industry; and on sociological ideas such as Callon's actor-networks to analyse this creation of market space by innovative firms. However, Green emphasizes the role of the firm or business enterprise as the central institution through which new products emerge as a result of its organized activities, and to which the sociological analyses have tended to pay too little attention (though Callon in chapter 4 describes the firm as an example of an actor-network). Even networks themselves have to some extent become institutionalized in joint ventures and inter-firm alliances. It is therefore firms' perceptions and interpretations of what other actors "do, mean and want — or can be persuaded to do, mean and want" — which are critical to technological shaping, and provide the social influences on artefact production. This is a point also taken up later by Vergragt *et al.*

Johan Schot points out in chapter 8 that past discussions of the interaction of technological and societal developments have rarely managed to conceptualize or unravel the nature of that interaction; they have tended merely to declare technological and socio-economic features to be complementary, or have provided a balance sheet of influential factors. Constructive technology assessment, developed in the Netherlands by Schot himself, Rip and others, represents an attempt to analyse the possible steering of technology in socially desirable directions, using a combination of analytical tools and traditions, including neo-Schumpeterian economics, social constructivism, the actor-network approach, and the systems approach.

Schot proposes a "quasi-evolutionary" model which is a compromise between the neo-Schumpeterian evolutionary model of variation and selection (also Metcalfe and Boden's starting point), and the insistence by sociologists of the influence of technology on the simultaneous shaping of content and context, instead of the existence of an independent selection environment. In his model, both content and context evolve, while variation and selection are linked by actors such as firms. Firms may adjust their search procedures so as to anticipate selection; there may be institutional links between the variation generation and the selection processes; and environmental requirements can be translated into criteria and specifications for technological development. This allows the selection environment to influence the generation of variety, and allows for the selection environment to influence variety without always being changed itself. Schot then discusses the development of clean technologies in the light of his model.

A focus on cultural and cognitive influences on decision making and corporate behaviour is an important feature of evolutionary economics, and an area where sociological approaches influence economic ones. Several chapters of this book have in different ways used concepts of culture and cognition. In chapter 9 Andrew Tylecote and Istemi Demirag examine short-term pressures on British manufacturing firms, their external and internal sources, and their structural and cultural causes. Short-term pressures have a profound effect on the rate and pattern of technological change. Cultural factors are found to influence strongly the behaviour of financial institutions and the stock market, and therefore the performance pressures they in turn impose on firms. For instance, German and Japanese investors are less likely to exert short-term pressure than their British or American counterparts (see also MacKenzie's chapter). Thus inadequate performance by a firm is likely to trigger the response of selling shares in the UK or US, but one of investors acquiring a "voice" in the running of the firm, in order to improve its performance, in Germany or Japan. In common with several other authors in this book, Tylecote and Demirag stress the importance of perceptions, values and preferences, which influence the pattern of search for information and the evaluation of that information. In addition to national cultural patterns, short-term pressures are likely to be more severe in an economic downturn, and to vary considerably with the level of technology in the firm or industry.

In chapter 10, Philip Vergragt, Peter Groenewegen and Karel Mulder use the concept of "dimensions of the innovative decision space", and analyse the cases of five high-performance fibres developed in several large industrial corporations, to explore the potential interaction of economic and sociological analyses of company strategy. There are three of these "dimensions": technological development, organizational structure, and the environment of the firm; which is not dissimilar to Metcalfe and Boden's theme of strategy emerging at the intersection of three types of phenomena. Vergragt *et al.* argue that a firm's strategy derives from its perception of the opportunities and constraints in these three dimensions, and that its innovation performance will be accomplished as a result of a process of choice-making in the context of this strategy.

They suggest that the mobilization of networks of communication and support both inside the firm, and with external financial and government institutions, customers, end users, suppliers, trade unions, environmentalists and others appropriate to the technology, and which shape the technology (usually analysed in sociological terms), are as important as the investment of resources (as analysed in economic terms). The search for a strong fibre (in terms of its physical characteristics) was translated into the search for a "strong" fibre (meaning a competitively strong product in the market), in the way Callon and others have used the term "translation". Economic motives were transformed by social actors in the firm into other arguments, interests, goals and motives, more susceptible to sociological analysis. Perception of the market, the position of competitors, and the firm's own strength were more important in strategic choice than "reality" (however measured or defined).

MacKenzie's discussion of uncertainty is relevant to the findings of Vergragt *et al.*, that the fibres that could be called successful innovations were not as successful as their innovators forecast, nor were they developed according to plan. Success was dependent on the exploitation of markets quite different from those initially identified in one case, and on the unexpected problems encountered by a competitor in another. The fibres that were commercial failures were not technological failures, or were not wholly technological failures. They may have been marginal to the firms' core businesses, or the R and D activity was not well integrated with the rest of the business activities, or the failure of relationships with other firms or organizations was a crucial blow.

CONCLUSION

The chapters which follow cover a wide range of theoretical positions and use a rich variety of empirical material to illustrate those positions. In our view they all reinforce, either in whole or in part, the broad diagnosis of convergence offered in this chapter. This convergence is first a convergence in terms of object of study, and second, but to a lesser degree as yet, in terms of analytical tools. Common concepts are used at times, although they have different meanings in the two disciplines. Thus there is clearly not a direct correspondence between, to take but one example, Metcalfe and Boden's "strategic paradigms" and Callon's "convergent networks". Both imply the intertemporal stability or invariance of at least some of the properties of the systems studied. However, in spite of the differences in meaning, they are at least not *incompatible* with each other, and this is a parallelism not to be taken lightly. Indeed, it could be argued *à la* Callon that this convergence is itself the creation of a new network, as the participants translate one another and eagerly accept the translations of others. Whether such converging trends will continue and will lead to a unified set of concepts to be used on both sides of the economics/sociology disciplinary divide, or whether this temporary convergence is a prelude to further divergence, we cannot anticipate. We certainly think it desirable for the interface between economics and sociology in the fields of technical change and of firm strategy to be kept as open as possible. To this extent we confidently expect these connections to be further explored and developed in the future.

References

Abernathy, W.J. and Utterback, J.M. (1975) A Dynamic Model of Process and Product Innovation, *Omega*, 3 (6) 639–56.

Abernathy, W.J. and Utterback, J.M. (1978) Patterns of industrial innovation, *Technology Review*, (June/July), 41–7.

Arthur, W.B. (1983) "Competing technologies and lock-in by historical events: the dynamics of allocation under increasing returns", International Institute for Applied Systems Analysis, Paper WP-83-90, Laxenburg, Austria.

Arthur, W.B. (1988) Competing technologies: an overview, in *Technical Change and Economic Theory* (Eds G. Dosi, *et al.*), Pinter, London.

Arthur, W.B. (1989) Competing technologies, increasing returns, and lock-in by historical events, *The Economic Journal*, 99 116–31.

Barnes, B. (1972) (ed.), *Sociology of Science*, Penguin Books, Harmondsworth.

Bijker, W.E., Hughes, T.P. and Pinch, T. (1986) *The Social Construction of Technological Systems: New Directions in the Sociology and History of Technology*, MIT Press, Cambridge, MA.

Braverman, H. (1974) *Labor and Monopoly Capital*, Monthly Review Press, New York.

Callon, M., Law, J. and Rip, A. (1986) *Mapping the Dynamics of Science and Technology*, Macmillan, London.

Callon, M. (1986) Some elements of a sociology of translation: domestication of the scallops and the fishermen of St. Brieuc Bay, in *Power, Action and Belief*, (ed. J. Law), Routledge and Kegan Paul, London.

Chandler, A. (1962) *Strategy and Structure*, MIT Press, Cambridge, MA.

Chandler, A. (1977) *The Visible Hand*, Harvard University Press, Cambridge, MA.

Child, J. (1972) Organisation Structure, Environment and Performance: The Role of Strategic Choice, *Sociology*, 6, 1–22.

Coombs, R., Saviotti, P. and Walsh, V. (1987) *Economics and Technological Change*, Macmillan, London.

Cyert, R.M. and March, J.G. (1963) *A Behavioural Theory of the Firm*, Prentice Hall, Englewood Cliffs, NJ.

Dosi, G. (1982) Technological Paradigms and Technological Trajectories: a Suggested Interpretation of the Determinants and Directions of Technical Change, *Research Policy*, **11**.

Dosi, G., Freeman, C., Nelson, R., Soete, L. and Silverberg, G. (1988) *Technical Change and Economic Theory*, Pinter, London.

Emery, F.E. and Trist, E.L. (1960) Socio-technical systems, in *Management Science, Models and Techniques*, (Eds C.W. Churchman and E.L. Verhulst), Vol. 2, Pergamon Press, London.

Emery, F.E. and Trist, E.L. (1965) The causal texture of organisational environments, *Human Relations*, **18**, 21–32.

Farmer, M. and Matthews, M. (1990) "Cultural Difference and Subjective Rationality: Where Sociology Connects with the Economics of Technical Choice", paper to the Conference on Firm Strategy and Technical Change: Microeconomics or Microsociology?, Manchester.

Georgescu Roegen, N. (1971) *The Entropy Law and the Economic Process*, Harvard University Press, Cambridge, MA.

Hodgson, G.M. (1991) Evolution and intention in economic theory, in *Evolutionary Theories of Economic and Technological Change: Present State and Future Prospects* (Eds P.P. Saviotti and J.S. Metcalfe), Harwood Publishers, London.

Johnson, G. and Scholes, K. (1984) *Exploring Corporate Strategy*, Macmillan, London.

Kay, N. (1982) *The Emergent Firm*, Macmillan, London.

Knights, D. and Morgan, G. (1990) Corporate Strategy, Organisations, and Subjectivity: A Critique, *Organisation Studies* **12** (2), 251–73.

Mangematin, V. and Callon, M. (1991) "Technological competition, strategies of the firms and the choice of the first users: the case of road guidance technologies", presented at the Colloquium on Management of Technology, Conference des Grandes Ecoles, Paris, May 27–9.

Metcalfe, J.S. (1981) Impulse and diffusion in the study of technical change, *Futures*, **13**.

Metcalfe, J.S. (1988) The diffusion of innovation, in *Technical Change and Economic theory*, (Eds G. Dosi, C. Freeman, R. Nelson, L. Soete, and G. Silverberg), Pinter, London.

Metcalfe, J.S. and Gibbons, M. (1989) Technology, variety and organisation: a systematic perspective on the competitive process, *Research on Technological Innovation, Management and Policy*, **4** 153–93.

Mintzberg, H. (1978) Patterns in Strategy Formation, *Management Science*, **14**, 934–48.

Moss, S. (1981) *An Economic Theory of Business Strategy*, Martin Robertson, Oxford.

Nelson, R. and Winter, S. (1977) In Search of Useful Theory of Innovation, *Research Policy*, **6**, 36–76.

Nelson, R. and Winter, S. (1982) *An Evolutionary Theory of Economic Change*, Harvard University Press, Cambridge, MA.

Nelson, R. (1990) Capitalism as an engine of progress, *Research Policy*, **19** 193–214.

Pavitt, K. (1990) "Key Characteristics of the Large Innovating Firm", mimeo, Science Policy Research Unit, Sussex University.

Penrose, E. (1959) *The Theory of the Growth of the Firm*, Blackwell, Oxford.

Pettigrew, A. (1987) Context and Action in the Transformation of the Firm, *Journal of Management Studies* **24** (6), 649–69.
Polanyi, M. (1962) *Personal Knowledge: Towards a Post-Critical Philosophy*, Harper Torchbooks, New York.
Porter, M.E. (1980) *Competitive Strategy*, The Free Press, New York.
Rosenberg, N. (1982) *Inside the Black Box: Technology and Economics*, Cambridge University Press, Cambridge.
Sahal, D. (1981a) Alternative conceptions of technology, *Research Policy*, **10**, 2–24.
Sahal, D. (1981b) *Patterns of Technological Innovation*, Addison-Wesley, Reading, MA.
Saviotti, P.P. and Metcalfe, J.S. (1984) A Theoretical Approach to the Construction of Technological Output Indicators, *Research Policy*, **13** 141–51.
Saviotti, P.P. (1988) "A characteristics approach to technological evolution and competition", Presented at the Conference on Recent Trends in the Economics of Technological Change" Manchester, 21–2 March.
Saviotti, P.P. (1991) "Technological evolution, variety and competition", presented at the Conference des Grandes Ecoles, Paris, 27–9 May.
Saviotti, P.P. and Metcalfe, J.S. (eds) (1991) *Evolutionary Theories of Economic and Technological Change: Present State and Future Prospects*, Harwood Publishers, London.
Schumpeter, J. (1934, original edition 1912), *The Theory of Economic Development*, Harvard University Press, Cambridge, MA.
Schumpeter, J. (1943) *Capitalism, Socialism and Democracy*, Harvard University Press, Cambridge, MA.
Shackle, G.L.S. (1955) *Uncertainty in Economics*, Cambridge University Press, Cambridge.
Shapiro, C. (1989) The theory of business strategy, *Rand Journal of Economics*.
Simon, H.A. (1969, new edition 1981) The natural and the artificial world, in *The Sciences of the Artificial*, (Ed. H.A. Simon) MIT Press, Cambridge, MA.
Teece, D.J. (1981) The market for know-how and the efficient international transfer of technology, *Proceedings of the National Academy of Political and Social Science*, **458**, 81–96.
Teece, D.J. (1986) Profiting from technological innovation, *Research Policy*, **15** 285–305.
Teece, D.J., Pisano, G. and Shuen, A. (1990) "Firm capabilities, resources and the concept of strategy", CCC Working paper No. 90–8, Center for Research In Management, University of California at Berkeley.
Williamson, O.E. (1975) *Markets and Hierarchies*, Free Press, New York.
Williamson, O.E. (1985) *The Economic Institutions of Capitalism*, Free Press, New York.
Woodward, J. (1965) *Industrial Organisation: Theory and Practice*, Oxford University Press, Oxford.

2

ECONOMIC AND SOCIOLOGICAL EXPLANATION OF TECHNICAL CHANGE

Donald MacKenzie

INTRODUCTION[1]

This chapter seeks to identify tools to overcome the cleavage between economic and sociological analyses of technological change. It draws on the tradition of "alternative economics" deriving from Herbert Simon. A more implicit debt is to Marx's critique of political economy, while an explicit, but of necessity highly tentative, attempt is made to argue that the sociology of scientific knowledge might be brought to bear on the economist's discussion of the unmeasurable uncertainty (rather than quantifiable risk) of technological change.

Rejecting neoclassical economics because of the incoherence of the notion of profit maximization, it examines the promising "alternative economics" associated with Simon, Nelson and Winter. It suggests that Nelson, Winter and Dosi's notion of "technological trajectory" contains a crucial ambiguity, and could productively be reconceptualized.

The paper calls for "ethnoaccountancy" of technological change. It ends by arguing that the sociology of scientific knowledge has much to contribute to analysis of uncertainty, a topic central to technological change.

I am painfully aware of many places where the chapter strays into areas where I am ignorant. It may well be that answers exist to the questions I ask, that there is relevant literature of which I am unaware. It may be that, as a sociologist, I have misunderstood what economists mean. In some places I suspect, though am not certain, that I am calling for the bringing of coals to Newcastle. If any of these possibilities is so, I would be most grateful for pardon. Unless we take the risk of revealing our ignorance, interdisciplinary bridges will not be built.

In studies of technology, the gap between economic and sociological explanation is pervasive. Economic analysis is often based upon assumptions sociologists regard as absurd, while sociological writing often almost ignores the dimension of cost and profit in its subject matter. Though there are thinkers who have provided rich resources for transcending the gap — despite their considerable differences, Karl Marx and Herbert Simon are the two central ones — it is far more common to find economic and sociological studies, even of the same topic, existing in separate conceptual universes.[2]

The first secton of the chapter contrasts neoclassical economics, particularly its assumption of profit maximization, with the alernative economics associated with Simon and more recently developed most fully by Richard Nelson and Sidney Winter. I then go on to discuss possible applications of that alternative view, to a false dichotomy sometimes found in labour process studies, to pricing behaviour in the computing industry and to the setting of R and D budgets.

Next follows an examination of the idea of "technological trajectory" or "natural trajectory" of technology to be found in the work of Nelson and Winter and other recent contributors to the economics of technology. I argue that although persistent patterns of technological change do exist, there is a crucial ambiguity in their description as "natural", and that a different understanding of them would help bridge the gap between economic and sociological explanation.

The following section discusses another way of bridging the gap, one again loosely in the tradition of Simon, but in practice little pursued: the "ethnoaccountancy" of technological change, the empirical study of how people actually reckon financially about technology (as distinct from how economic theory suggests they should reckon).

Finally, I turn to the topic of uncertainty and the construction of the economic. Despite their "thing-like" character, economic relations are never wholly self-sustaining and self-explaining. While this point is normally argued in the large (Marx justifies it by an examination of the evolution of capitalism), technological innovation demonstrates it in the smaller scale. As is well known, the inherent uncertainty of radical innovation makes economic calculation applicable only *ex post*, not *ex ante* — once networks have stabilized, not before they stabilize. This makes radical innovation a problem for (orthodox) economics, but points, I argue, to the relevance here of the sociology of scientific knowledge.

NEOCLASSICAL AND ALTERNATIVE ECONOMICS

It is convenient to begin with our feet firmly on the economic side of the gap. The neoclassical economics of production technology is crystalline in its explanations. Although the full neoclassical structure is dauntingly complex, its central pivot is simple and clear. Firms choose production technology so as to maximize their rate of profit.

Unfortunately, that clarity is purchased at too high a price. The notion of maximization, at the heart of the neoclassical structure, is incoherent, at least as a description of how firms do, or even could, behave. Perhaps the most cogent statement of why this is so comes from Sidney Winter:

> It does not pay, in terms of viability or of realized profits, to pay a price for information on unchanging aspects of the environment. It does not pay to review constantly decisions which require no review. These precepts do not imply merely that information costs must be considered in the definition of profits. For without observing the environment, or reviewing the decision, there is no way of knowing whether the environment is changing or the decision requires review. It might be argued that a determined profit maximizer would adopt the organization form which calls for observing those things that it is profitable to observe at the times when it is profitable to observe them: the simple reply is that this choice of a profit maximizing information structure itself requires information, and it is not apparent how the aspiring profit maximizer acquires this information, or what guarantees that he does not pay an excessive price for it.[3]

This critique of neoclassical economics draws most importantly upon the work of Herbert Simon, and has been elaborated by Winter, his collaborator Richard Nelson, and other economists. Its logic seems inescapable,[4] though I confess to having been surprised by economist colleagues who, despite proclaimed hostility to neoclassical economics, smuggle maximizing assumptions into their explanations.

Simon and his intellectual descendants of course do more than highlight the central incoherence haunting neoclassical economics' formidable apparatus of production functions, isoquants and the like. They provide a different vision of economic activity. In this alternative economics, actors follow routines, recipes, and rules of thumb, while monitoring a small number of feedback variables. While the values of these are satisfactory — "satisficing" is, of course, Simon's famous replacement for maximizing — routines continue to be followed. Only if they become unsatisfactory will they be reviewed, but the review will not be an unconstrained evaluation of the full universe of alternatives in search of the best. It will be a local search, given direction by the perceived problem in need of remedy and using heuristics, which are rather like routines for searching.

This intellectual tool-kit offers a bridge towards the sociological, as conventionally understood. Routines can be entrenched for a variety of organizational reasons, and different parts of a firm will typically follow different routines and different heuristics of search. Since, in this perspective, there is no longer any ultimate arbiter of routines (such as profit maximization), firms become political coalitions, rather than unitary rational decision-makers. The actual behaviour of a firm may represent a compromise between different, potentially contending, courses of action.[5]

Intra-firm processes are not, of course, ultimately insulated from what goes on outside the firm. That outside is a "selection environment", favouring certain routines and not others. Nelson and Winter, especially, draw an explicit parallel with evolutionary biology, seeing routines as akin to genes, selected for or against by their environment. This environment is not just "the market", but includes other institutional structures as well. It is not necessarily, or even generally, stable, nor is it simply external and "given". One particular firm may be able to alter its environment only slightly (although some patently alter it more than slightly), but the behaviour of the ensemble of firms is in large part what constitutes the environment.[6]

This "alternative economics" promotes a subtle change in ways of thinking, even in areas where its relevance is not apparent. Take, for example, David Noble's justifiably celebrated, empirically rich, study of the automation of machine tools in the USA. Noble frames his most general conclusion in terms of a dichotomy between profit and capitalists' control over the workforce:

> It is a common confusion, especially on the part of those trained in or unduly influenced by formal economics (liberal and Marxist alike), that capitalism is a system of profit-motivated, efficient production. This is not true, nor has it ever been. If the drive to maximize profits, through private ownership and control over the process of production, has served historically as the primary means of capitalist development, it has never been the end of that development. The goal has always been domination (and the power and privileges that go with it) and the preservation of domination.[7]

This analytical prioritization of the sociological[8] over the economic cannot be correct: a firm or industrial sector that pursued control at the expense of profit would shrink or die. Much of the industrial sector in the USA studied by Noble did indeed suffer this fate, in the period subsequent to that examined, at the hands of the Japanese machine-tool manufacturers, equally capitalist, but, in their organizational and technological choices, less concerned with control over the workforce.

What leads Noble into posing the false dichotomy of the economic and the sociological, I would suggest, is the echo of neoclassical economics. The alternative economics shows how to make analytical sense of capitalists who are profit oriented (as any sensible view of capitalists must surely see them), without being profit maximizers. The urge to capitalist control is not an overarching imperative of domination, ultimately overriding the profit motive, but an "heuristic"[9] with deep roots in the antagonistic social relations of capitalist society. When facing technological choices, American engineers and managers, in the period studied by Noble, often simplified production technology decisions by relying on an entrenched preference for technological solutions that undercut the position of manual labour. Noble quotes a 1968 article by Michael Piore, based on an extensive survey of engineers:

> Virtually without exception, the engineers distrusted hourly labor and admitted a tendency to substitute capital whenever they had the discretion to do so. As one engineer explained, "if the cost comparison favored labor but we were close, I would mechanize anyway."[10]

Any significant technological change (such as the automation of machine tools) involves deep uncertainty as to future costs and therefore profits, uncertainty far more profound than the quotation from Piore's work implies. Relying on simple heuristics to make decisions under such circumstances does not demonstrate indifference to profit: there is simply no completely rational, assuredly profit-maximizing, way of proceeding open to those involved.[11] Thinking in terms of heuristics rather than imperatives might also open up a subtly different set of research questions about the interaction of engineers' culture with the social relations (including the economic relations) of the workplace, and about the different heuristics found under different circumstances (including different national circumstances).

Existing attempts to give empirical content to the ideas of the alternative economics have, however, naturally been more traditionally "economic" than that sort of investigation. Pricing behaviour is perhaps the most obvious example.[12] Prices do typically seem to be set according to simple, predictable rules of thumb. Even in the sophisticated US computer industry, what may be the basic rule is startlingly simple: set selling price at three times manufacturing cost.[13] Of course, much more elaborate sets of procedures have evolved (along with the specialist function of pricing manager). These procedures, however, still seem likely to be comprehensible in the terms of the alternative economics, and indeed open to research (although, probably through ignorance, I know of no published study of them). Cray Research, for example, sets its supercomputer prices according to a well-defined financial model, in which the relevant rule is that 40 per cent of the proceeds of a sale should cover manufacturing cost plus some parts of field maintenance, leaving 60 per cent overhead.[14] Discounting and different ways of determining manufacturing cost make such rules, even if simple in form, flexible in application, but I would speculate (a) that understanding them is an essential part of understanding the computer industry, and (b) that they are by no means accidental, but (like the control heuristic) have deep roots. It would, for example, be fascinating to compare Japanese and American computer industry pricing. There is certainly some reason to think that, in general, Japanese prices may be set according to heuristics quite different from those that appear prevalent in the US.[15] If this is correct for computing, it is unlikely to be an accidental difference, but will be connected to considerable differences in the organizational, financial and cultural circumstances of the two computer industries.

Similarly, it has frequently been asserted that large firms determine total R and D budgets by relatively straightforward rules of thumb.[16] At Cray Research, for example, the R and D budget is set at 15 per cent of total revenue.[17] On the other hand, some recent British evidence suggests that matters are not always that straightforward,[18] and that there seem likely to be many other complications, such as the significance of the definition of expenditure as R and D for taxation and for perception of a firm's future prospects. Here too, however, is an area where empirical investigation inspired by the alternative economics might be most interesting.[19]

TRAJECTORIES

What, however, of the *content* of R and D, rather than its quantity? Perhaps the most distinctive contribution in this area of recent work within the tradition of "alternative economics" is the notion of technological trajectory, or "natural trajectory" of technology.[20]

That there is a real phenomenon to be addressed is clear. Technological change does show persistent patterns: for example, the increasing mechanization of manual operations, the growing miniaturization of microelectronic components, the increasing speed of computer operations. Some of these patterns are indeed so precise as to take a regular quantitative form. For example, "Moore's Law" concerning the annual doubling of the number of components on state-of-the-art microchips was formulated in 1964, and has held remarkably well (with, at most, only a gradual increase in doubling time in recent years) from the first planar-process transistor in 1959, to the present day.[21]

The problem, of course, is how such persistent patterns of techno-logical change are to be explained. "Natural" is here a dangerously ambiguous term. One meaning of the term is "what is taken to follow as a matter of course", what people unselfconsciously set out to do, without external prompting. That is the sense of "natural" in the following passage from Nelson and Winter:

> ... the result of today's searches is both a successful new technology and a natural starting place for the searches of tomorrow. There is a "neighborhood" concept of a quite natural variety. It makes sense to look for a new drug "similar to" but possibly better than the one that was discovered yesterday. One can think of varying a few elements in the design of yesterday's successful new aircraft, trying to solve problems that still exist in the design or that were evaded through compromise.[22]

The trouble, however, is that "natural" has quite another meaning, connoting what is produced by, or according to, nature; not the work of humans. That other meaning might not be troublesome, did it not resonate with a possible interpretation of the mechanical[23] metaphor of "trajectory". If I throw a stone, I as human agent give it initial direction. Thereafter, its trajectory is influenced by physical forces alone. The notion of "technological trajectory" can thus very easily be taken to mean that once technological change is initially set on a given path (for example, by the selection of a particular paradigm), its development is then determined by technical forces.

If Nelson and Winter incline to the first meaning of natural, Giovanni Dosi — whose adoption of the notion of trajectory has been at least equally influential — can sometimes, if not always,[24] be read as embracing the second. To take two examples:

> "Normal" technical progress maintains a momentum of its own which defines the broad orientation of the innovative activities.

> Once a path has been selected and established, it shows a momentum of its own.[25]

A persistent pattern of technological change does indeed possess momentum, but never momentum of its own. Historical case-study evidence can be brought to bear to show this (such as Tom Hughes's study, rich in insights, of the trajectory of hydrogenation chemistry), as can the actor-network theory of Michel Callon, Bruno Latour, John Law and colleagues.[26] I shall, however, argue the point rather differently, drawing on an aspect of trajectories that is obvious but, surprisingly, seems not to have been developed in the literature on the concept.[27]

That is, that a technological trajectory can be seen as a self-fulfilling prophecy. Persistent patterns of technological change are persistent, in part, because technologists and others believe they will be persistent.

Take, for example, the persistent increase in the speed of computer calculation. At any point in time, there seems to be a reasonably consensual estimate of the likely rate of increase in supercomputer speed: that it will, for example, increase by a factor of ten every five years.[28] This kind of estimate is drawn upon by supercomputer designers to help them judge how fast their next machine must be if it is to compete with those of their competitors, and thus the estimate is an important factor shaping supercomputer design. Thus the designer of the ETA[10] supercomputer told me that he determined the degree of parallelism of this machine's architecture by deciding that it must be ten times faster than its Cyber 205 predecessor. Consulting an expert on microchip technology, he found that the likely speed-up in basic chips was of the order of fourfold. The degree of parallelism was then determined by the need to obtain the remaining factor of 2.5 by using multiple processors.[29]

While I have not yet been able to interview Seymour Cray or the designers of the Japanese supercomputers, such evidence as exists does suggest similar processes of reasoning in the rest of mainstream supercomputing (massively parallel architectures and mini-supercomputers are different). If possible, speed has been increased by the amount assumed necessary by using faster components, while preserving the same architecture, and thus diminishing risks and reducing problems of compatibility with existing machines. When sufficiently faster components have not been thought available, architectures have been altered to gain increased speed by various forms of parallelism.

The prophecy not just of increased speed, but of a specific rate of increase, has thus been self-fulfilling. While it has clearly served as an incentive to technological ambition, less obviously, it has also served to limit such ambition. Why, the reader may ask, do designers satisfice rather than seek to optimize? Why do they not design the fastest possible computer (indeed the latter is what they, and particularly Seymour Cray, are often portrayed as doing)? The general difficulties of the concept of optimization aside, the specific problems are risk and cost. By general consensus, the greater the speed goal, the greater the risk of technological failure, and the greater the ultimate cost of the machine. Though supercomputer customers are well-heeled, there is still assumed to be a band of "plausible" supercomputer cost (currently roughly $10 million to $30 million). If designers do not moderate their ambitions to take risk and cost into account, their managers and financiers will.[30] The assumed rate of speed helps as a yardstick to know what is an appropriately realistic level of ambition.

In this case, all those involved are agreed that increased speed is desirable. Similarly, all those involved with chip design seem to assume that, other things being equal, increased component counts are desirable. Trajectories are self-fulfilling prophecies, however, even when that is not so. Take the mechanization of processes previously done by hand. Though analysed as a natural trajectory by Nelson and Winter,[31] it has of course often seemed neither natural nor desirable to those involved, particularly to workers fearing for their jobs or skills, but sometimes also to managements disliking change, investment and uncertainty. A powerful argument for mechanization, however, has been the assumption that other firms, and other countries, will mechanize, and that a firm that does not will go out of business. Increasing missile accuracy is a similar, if simpler, case: those who have felt it undesirable (because it might ultimately lead to a nuclear "first strike" on an opponent's forces) have often felt unable to oppose the process, because they have assumed it to be inevitable, and, specifically, not stoppable by arms control agreements. Their consequent failure to oppose it has been one factor making it possible.

The nature of the technological trajectory as self-fulfilling prophecy can be expressed in the language both of economics and of sociology. As an economist would put it, expectations are an irreducible aspect of patterns of technological change. The work of Brian Arthur and Paul David is of relevance here, although it has, to my knowledge, to date concerned either/or choices of technique or standard, rather than the cumulative, sequential decisions that make up a trajectory. In an amusing and insightful discussion of the almost universal adoption of the inferior qwerty keyboard, David writes:

> Intuition suggests that if choices were made in a forward-looking way, rather than myopically on the basis of comparisons among currently prevailing costs of different systems, the final outcome could be influenced strongly by the expectations that investors in system components — whether specific touch-typing skills or typewriters — came to hold regarding the decisions that would be made by the other agents. A particular system could triumph over rivals merely because the purchasers of the software (and/or the hardware) expected that it

would do so. *This intuition seems to be supported by recent formal analyses of markets where purchasers of rival products benefit from externalities conditional upon the size of the compatible system or "network" with which they thereby become joined.*[32]

Actors' expectations of the technological future are part of what make a particular future, rather than other possible futures, real. With hindsight, the path actually taken may indeed look natural, indicated by the very nature of the physical world. But Brian Arthur's "non-ergodic", path-dependent, models of adoption processes are vitally helpful in reminding us of ways in which technologies devoid of initial intrinsic superiority can rapidly become irreversibly superior in practice, through the very process of adoption.[33]

The sociological way of expressing essentially the same point is to say that a technological trajectory is an "institution". Like any institution it is sustained not through any internal logic, or intrinsic superiority to other institutions, but because of the interests that develop in its continuance and the belief that it will continue. Its continuance becomes embedded in actors' frameworks of calculation and routine behaviour, and it continues because it is thus embedded. Although it is intensely problematic to see institutions as natural in the sense of corresponding to nature (although that is how they are frequently legitimated), institutions do of course often become natural in the sense of being unselfconsciously taken-for-granted. The recent sociological work that is of greatest relevance here is that of Barry Barnes, who has argued that self-fulfilling prophecy should be seen not as a pathological form of inference (as it often was in earlier sociological discussions), but as the basis of all social institutions, including the pervasive phenomenon of power.[34]

My claim is not the idealist one that all prophecies are self-fulfilling. Many widely-held technological predictions prove false: a good example is the assumption of ever-increasing speed in civil air transport.[35] Not all patterns of technological change can be institutionalized, and it would be foolish to deny that the characteristics of the material world, of Callon and Latour's "non-human actors", play a part in determining the patterns that do become institutionalized. One reason for the attractiveness of the notion of a natural trajectory to alternative economics has been that it has been reacting not against technological determinism (as has much of the sociology of technology), but against a view of technology as an entirely plastic entity shaped at will by the all-knowing hands of market forces.[36] I entirely sympathize with the instinct that the world of technology cannot be shaped at will, whether by markets or societies.

The risk, however, of expressing that valid instinct in the notion of natural trajectory is that it may actually deaden intellectual curiosity about the causes of persistence in patterns of technological change. Although I am certain this is not intended by its proponents, the term has an unhappy resonance with widespread (if implicit) prejudices about the proper sphere of social science analysis of technology, prejudices that shut off particular lines of enquiry. Let me give just one example. There is wide agreement that we are witnessing an information technology "revolution", or change of "technoeconomic paradigm" based around information and communication technologies. Of key importance to that revolution, or new paradigm, is, by general agreement, microchip technology, and its "Moore's Law" pattern of development: "clearly perceived low and rapidly falling relative cost", "apparently almost unlimited availability of supply over long periods" and "clear potential for . . . use or incorporation . . . in many products and processes throughout the economy".[37]

Yet in all the plethora of economic and sociological studies of information technology, there is not a single piece of published research — and I hope I do not write from ignorance here — on the determinants of this central Moore's Law pattern.[38] Explicitly or implicitly, it is taken to be a natural trajectory whose effects economists and sociologists may study but whose causes lie outside their ambit. In Dosi's work on semiconductors, for example, Moore's Law is described as "almost a 'natural law' of the industry", a factor shaping technical progress, but not one whose shaping is itself to be investigated.[39] Until such a study of Moore's Law is done, we cannot say precisely what intellectual opportunities are being missed, but it is unlikely that they are negligible.[40]

ETHNOACCOUNTANCY

A revised understanding of persistent patterns of technological change offers one potential bridge over the gap between economic and sociological explanations of technical change. Another potential bridge I would call "ethnoaccountancy". I intend the term to be analogous to ethnomusicology, ethnobotany, or ethnomethodology. Just as ethnobotany is the study of the way societies classify plants, a study that should not be structured by our perceptions of the validity of these classifications, ethnoaccountancy should be the study of how people do their financial reckoning, irrespective of our perceptions of the adequacy of that reckoning, and of the occupational labels attached to those involved.

Ethnoaccountancy has not been a traditional concern of writers within the discipline of accounting. Their concern naturally has been with how accountancy ought to be practised, rather than with how it actually is.[41] Although studies of the latter have been much more common over the past decade (see, for example, the pages of the journal *Accounting, Organizations and Society*), there has still been little systematic study by accountancy researchers of the ethnoaccountancy of technological change. Sociologists, generally, have not been interested in ethno-accountancy, again at least until very recently.[42] Compare, for example, the enormous bulk of the sociology of medicine with the almost non-existent sociology of accountancy.[43] Given that the latter profession could arguably be as important to the modern world as the former, it is difficult not to suspect that sociologists have been influenced by accountancy's general image as a field which may be remunerative, but which is deeply boring.

It is somewhat more surprising that economists have ignored the actual practices of accounting, but that appears to be the case. Nelson and Winter suggest a reason which, though tendentiously expressed, may be essentially correct:

> For orthodoxy, accounting procedures (along with all other aspects of actual decision processes) are a veil over the true phenomena of firm decision making, which are always rationally oriented to the data of the unknowable future. ... Thanks to orthodoxy's almost unqualified disdain for what it views as the epiphenomena of accounting practice, it may be possible to make great advances in the theoretical representation of firm behavior without any direct empirical research at all — all one needs is an elementary accounting book.[44]

Ethnoaccountancy most centrally concerns the category of "profit". As noted above, even if firms cannot maximize profit, it certainly makes sense to see them as being profit-oriented. But they can know their profits only through accounting practices. As these change, so does the meaning, for those involved, of profit. Alfred Chandler's *The Visible Hand*, for example, traces how accounting practices, and the definition of profit, changed as an inseparable part of the emergence of the modern business enterprise.[45] Unfortunately, Chandler clothes his informed analysis in teleological language — he describes an evolution towards correct accounting practice and a "precise" definition of profit[46] — and he does not directly tie the changes he documents to changing evaluations of technology.

The teleology has largely been corrected and the connection to technological change forged, albeit in a much more limited domain, by historian of technology Judith McGaw.[47] Though adequate for the purposes of those involved, accounting practice in the early nineteenth-century US papermaking industry, she notes, "hid capitalization" and highlighted labour costs, facilitating the process of the mechanization of manual tasks. Though others have not made the same connections she has, it is clear that the practices she documents were not restricted to the particular industry she discusses.[48]

The general issue of whether accounting practice highlights one particular class of cost, thus channelling innovation towards the reduction of that cost, is of considerable significance. Accounting practices which highlight labour costs might generally be expected to accelerate mechanization. They may, however, be a barrier to the introduction of capital-saving or energy-saving technologies, and many current information technology systems are regarded as having these advantages.

There is also fragmentary but intriguing evidence that the techniques of financial assessment of new technologies used in the UK and US may differ from those used in Japan. In effect, profit is defined differently. In the UK and US there is typically great reliance (for decision-making purposes, and also in rewarding managers) on what one critic calls

> ... financial performance measures, such as divisional profit, [which] give an illusion of objectivity and precision [but which] are relatively easy to manipulate in ways that do not enhance the long-term competitive position of the firm, and [which] become the focus of opportunistic behavior by divisional managers.[49]

Japanese management accounting, by contrast, is less concerned with financial measurement in this short-term sense. While Japanese firms are patently not indifferent to profit, and are of course legally constrained in how profit is calculated for purposes such as taxation, they seem much more flexible in the internal allocation of costs and definition of profit. Japanese firms "seem to use [management] accounting systems more to motivate employees to act in accordance with long-term manufacturing strategies than to provide senior management with precise data on costs, variances, and profits."[50]

UNCERTAINTY AND CLOSURE

Ethnoaccountancy is one aspect of the much larger topic we might call the construction of the economic. Economic phenomena such as prices, profits and markets are not just "there" — self-sustaining, self-explaining — but exist only to the extent that certain kinds of relations between people exist. This insight, simultaneously obvious and easy to forget, is perhaps Marx's most central contribution to our topic. Marx insisted that "capital is not a thing, but a social relation between persons which is mediated through things",[51] and devoted the final part of volume one of *Capital* to an analysis of the historical emergence of that particular way of mediating relations between persons. Implicit, too, in Marx's account is the reason why the insight is forgettable. It is not just that capitalism gives rise to a particular type of economic life. Under capitalism, aspects of social relations inseparable in previous forms of society (such as political power and economic relations) achieve a unique degree of separation, giving rise to the "thing-like" appearance of the economic.

One of the fascinations of technological change is that it turns the question of the construction of the economic from a general issue about capitalist society into a specific and unavoidable concern. The oft-noted unquantifiable uncertainty of technological change defies the calculative frameworks of economics. Chris Freeman, for example, compares attempts at formal evaluation of research and development projects to "tribal war-dances".[52] He is referring to participants' practices, but it is worth noting that the economists of technological change, in their search for an ancestor to whom to appeal, have often turned to Schumpeter, with his emphasis on the non-calculative aspects of economic activity, rather than any more orthodox predecessor.

The issue can usefully be rephrased in the terms of the actor-network theory of Callon, Latour and colleagues. Radical technological innovation requires the construction of a new actor-network:[53] indeed that is perhaps the best way of differentiating radical innovation from more incremental change. Only once a new network has successfully been stabilized does reliable economic calculation become possible.[54] Before it is established, other forms of action, and other forms of understanding, are needed.

Unstabilized networks are thus a problem for economics, at least orthodox economics. By comparison, however, their study has been the very life-blood of the sociology of scientific knowledge.[55] Scientific controversy, where the "interpretative flexibility" of scientific findings is made evident, has been its most fruitful area of empirical study; interpretative flexibility is the analogue of what the economists refer to as "uncertainty".[56] The weakness of the sociology of scientific knowledge has, rather, been in the study of "closure", the reduction of (in principle endless) interpretative flexibility, the resolution of controversy, the establishment of stable networks.

The economics of technological change and the sociology of scientific knowledge thus approach essentially the same topic — the creation of stable networks — from directly opposite points of view. I confess to what is perhaps a disciplinary bias as to how to proceed in this situation: using tools honed for stable networks to study instability seems to me likely to be less fruitful than using tools honed for instability to study stability.[57] Indeed, attempting the former is where, I would argue, the alternative economists have gone wrong in the concept of technological trajectory.

The latter path, using the tools developed in the study of instability, does, however, require a step backwards in research on technological change, a return to the "natural history"[58] of innovation of the 1960s and 1970s, but a return informed by new questions. Some of these questions — the empirical study of heuristics, the role of the self-fulfilling prophecy in persistent patterns of technological change, the ethnoaccountancy of technological change — have already been highlighted. The sociology of scientific knowledge suggests others. Most generally, we need to know more about the structure of the interpretative flexibility inherent in technological change, and about the ways that interpretative flexibility is reduced in practice. How, in economists' terminology, is uncertainty converted into risk?[59] How, for example, do participants judge whether they are attempting incremental or radical innovation?[60] What is the role of the testing of technologies (and analogies such as prototyping and benchmarking)?[61] How is technological change "packaged" for the purposes of management — in other words, how is a process, which from one perspective can be seen as inherently uncertain, presented as subject to rational control? What here is the role of project proposals, project reviews and milestones, of the different components of Freeman's "war-dances"? How is the boundary between the technical and the non-technical negotiated? What are the determinants of the credibility of technical, and non-technical, knowledge claims?

Even if we set aside the issue that technological change is not substantively the same as scientific change, the sociology of scientific knowledge cannot be looked to for theories or models that could be applied directly in seeking to answer questions such as this. That is not the way the field has developed. It is more a question of sensitivities, analogies, vocabularies and some intriguing (but far from proven) graphical and quantitative tools.[62]

Nevertheless, the parallels between closure in science and successful innovation in technology, and between interpretative flexibility and uncertainty, are strong enough to suggest that here there may be an important way forward for the study of technological change. In both, an apparently self-sustaining realm (of objective knowledge, of economic processes) emerges, but only as the end-product of a process involving much more than either reality or economic calculation. Understanding of the one should surely help develop understanding of the other.

CONCLUSION

I have argued that the alternative economics associated with Simon, Nelson, Winter and others is more plausible than neoclassical economics, with its incoherent notion of profit maximization. Ideas from the former tradition could help bridge the gap between the economic and sociological in fields where they have not (to my knowledge) been widely drawn upon, such as labour process studies. This alternative economics can also be applied fairly straightforwardly to matters such as pricing and firms' overall research and development budgets, although even in these areas recent empirical work seems surprisingly sparse.

The application of the alternative economics to the content of research and development is more difficult. The metaphor of "technological trajectory" can mislead. Persistent patterns of technological change do exist, but they should not be seen as "natural" in the sense of corresponding to nature. Nor do they have a momentum of their own. Expectations about the technological future are central to them: they have the form of self-fulfilling prophecies, or social institutions. Conceiving of persistent patterns in this way offers one way of bridging the gap between economic and sociological explanations of technological change.

Another way of bridging the gap is what I have called ethnoaccountancy. Studying how people actually do the financial reckoning of technological change would bring together the economist's essential concern for the financial aspects of innovation with the sociologist's equally justified empiricism. I have suggested that ethnoaccountancy would not be a marginal enterprise, rummaging through the boring details of economic activity, but ought to throw light on central questions such as the practical definition of profit and the relative rate of technological change in different historical and national contexts.

Finally, I have argued that because uncertainty, or non-stabilized networks, is central to technological change, the sociology of scientific knowledge, with its experience in the study of the essentially equivalent matter of interpretative flexibility, ought to be of relevance here. Scientists construct stable, irreversible developments in knowledge in a world where no knowledge possesses absolute warrant; out of potential chaos, they construct established science. Technologists, workers, users and managers construct successful innovations in a world where technological change involves inherent uncertainty; out of potential chaos, they construct a world in which economics is applicable.

Notes

1   The research reported on here was supported by the Economic and Social Research Council's Programme on Information and Communication Technologies. Earlier versions of this chapter were read as a paper to the conference on "Firm Strategy and Technical Change: Microeconomics or Microsociology?", Manchester, 27–8 September 1990, and to a meeting of the ESRC New Technologies and the Firm Initiative, Stirling, 6–7 February 1991. Thanks to Tom Burns, Chris Freeman, John Law, Ian Miles, Albert Richards, Steve Woolgar and both the above audiences for comments, and particularly to the neoclassical economists for their toleration of a sociologist's rudeness!

2   Compare, for example, the two pioneering monographs on domestic work and domestic technology: Ruth Schwartz Cowan (1983), *More Work for Mother: The Ironies of Household Technology from the Open Hearth to the Microwave* Basic Books, New York, and Jonathan Gershuny (1983), *Social Innovation and the Division of Labour* Oxford University Press, Oxford. Cowan's analysis rests fundamentally on value commitments. Even though he has since held a chair in sociology, Gershuny employs a typically economic model of rational, maximizing choice.

3   Sidney Winter, as quoted in Jon Elster (1983), *Explaining Technical Change: A Case Study in the Philosophy of Science*, Cambridge University Press, Cambridge, 139–40.

4   The classical counter-argument is to be found in Milton Friedman (1953), "The Methodology of Positive Economics", in *Essays in Positive Economics* (Ed. M. Friedman), University of Chicago Press, Chicago, 3–43, especially p. 22. According to this, even if firms do not consciously seek to maximize, a process akin to natural selection goes on. Firms that happen to hit on a maximizing or near-maximizing strategy will grow, while those that do not will shrink or fail, and therefore maximizing strategies will prevail, even if those who pursue them do so for reasons quite distinct from the knowledge that they are maximizing. If the environment in which firms were operating was unchanging, this defence would be perfectly plausible. The assumption of an unchanging, "given" environment is, however, far too unrealistic and restrictive, especially

when technological change is being considered. If the environment is changing dramatically, it is far from clear that there is a stable maximizing strategy towards which selection will move populations. Game-theoretic elaborations to the neoclassical framework help, because they can model the way one firm's action changes another firm's environment, but even they rely on a certain stability of framework.

5 See R.M. Cyert and J.G. March (1963), *A Behavioral Theory of the Firm*, Prentice Hall, Englewood Cliffs, NJ; Richard R. Nelson and Sidney G. Winter (1982), *An Evolutionary Theory of Economic Change*, Harvard University Press, Cambridge, MA, 107–12. Note the evident parallels with debates in political science over the explanation of national policy, especially in the fields of defence and foreign affairs, where a "realist" position, akin to neoclassical economics, has contended with a "bureaucratic politics" position which has argued that policy is outcome, rather than decision. Two classic discussions are Graham Allison (1971), *Essence of Decision: Explaining the Cuban Missile Crisis*, Little, Brown, Boston, and John D. Steinbruner (1974), *The Cybernetic Theory of Decision: New Dimensions of Political Analysis*, Princeton University Press, Princeton, NJ.

6 Nelson and Winter (1982), *An Evolutionary Theory*.

7 David Noble (1984), *Forces of Production: A Social History of Industrial Automation* Knopf, New York, 321.

8 Although there is no indication that it is a connection which Noble would wish to draw, it is worth noting that "domination" is, of course, a key category of Max Weber's sociology.

9 Since capitalist domination can take a range of different forms, of which direct control is only one, it would be more correct to talk of a set of heuristics. The discussion in the text, for the sake of simplicity, deals with direct control alone.

10 Michael Piore (1986), The Impact of the Labor Market upon the Design and Selection of Productive Techniques within the Manufacturing Plant, *Quarterly Journal of Economics*, 82, as quoted by Noble (1984), *Forces of Production*, 217n. For more recent evidence bearing upon the same issue, see Michael L. Dertouzos, Richard K. Lester, Robert N. Solow and the MIT Commission on Industrial Productivity (1989), *Made in America: Regaining the Productive Edge* MIT Press, Cambridge, MA.

11 As Noble is fully aware, there is indeed a further factor of great importance in the automation of machine tools in the USA — the very heavy involvement of the US Air Force, and thus the specific "ethnoaccountancy" (see below) of military technology.

12 A classic study of pricing behaviour in the "alternative" tradition is the account of pricing in a department store in Cyert and March (1963), *A Behavioural Theory*, chapter seven. See also Nelson and Winter (1982), *An Evolutionary Theory*, 410.

13 Interview with Neil Lincoln (formerly leading supercomputer designer at Control Data Corporation and ETA Systems), Minneapolis, MN, 3 April 1990.

My data on the prevalence of the 3:1 rule is strictly limited, and so what I say in the text can only be tentative.

14 Interview with John Rollwagen (Chairman, Cray Research, Inc.), Minneapolis, MN, 3 April 1990. The overhead had recently been reduced from 65 per cent to 60 per cent.

15 Toshiro Hiromoto (1988), Another Hidden Edge — Japanese Management Accounting, *Harvard Business Review*, **66** (4) (Jul./Aug.), 22–6.

16 See, e.g., Christopher Freeman (1982), *The Economics of Industrial Innovation* (second edition), Pinter, London, 163; Rod Coombs, Paolo Saviotti and Vivien Walsh (1987), *Economics and Technological Change* Macmillan, Basingstoke, 57.

17 Rollwagen interview.

18 R. Ball, R.E. Thomas and J. McGrath (1991), "A Survey of Relationships between Company Accounting and R and D Decisions in Smaller Firms", paper read to meeting of ESRC New Technologies and the Firm Initiative, Stirling, 6–7 February. The authors found that firms reported giving greater weight to the assessment of project costs and benefits, than to simpler determinants such as the previous year's R and D budget. It could be, however, that firms were reporting idealized versions of their practice.

19 See Nelson and Winter (1982), *An Evolutionary Theory*, 251–4. Again, though, I have not been able to trace recent empirical work. Nelson and Winter's book in fact gives only passing attention to pricing and R and D budgets, and concentrates on developing quantitative, long-term economic growth models. Though these are impressive (and appear empirically successful), the assumptions built into them are too simple, and what is being explained is too general, for them to be of direct relevance here. There is a brief and clear summary of this aspect of Nelson and Winter's work in Paul Stoneman (1983), *The Economic Analysis of Technological Change* Oxford University Press, Oxford, 184–5.

20 Richard Nelson and Sidney Winter (1977), in Search of Useful Theory of Innovation, *Research Policy*, **6**, 36–76, especially 56–60; Nelson and Winter (1982), *An Evolutionary Theory*, 255–62; Giovanni Dosi (1982), Technological Paradigms and Technological Trajectories: A Suggested Interpretation of the Determinants of Technical Change, *Research Policy*, **11**, 147–62. A more recent discussion is Giovanni Dosi (1988), The Nature of the Innovative Process, in *Technical Change and Economic Theory* (Eds Dosi et al.) Pinter, London, 221–38. The concept is now in the textbooks. See Coombs et al. (1987), *Economics and Technological Change*.

21 For "Moore's Law", named after Gordon E. Moore, Director of Research at Fairchild Semiconductor in 1964, see Robert N. Noyce (1977), Microelectronics, *Scientific American*, **237** (3) (September 1977), reprinted in Tom Forester (ed.) (1980), *The Microelectronics Revolution* Blackwell, Oxford, 29–41.

22 Nelson and Winter (1982), *An Evolutionary Theory*, 257.

23 It is perhaps of some significance that Nelson and Winter, whose overall project is framed in a biological metaphor, here (without discussion) change from biology to mechanics.

24   Thus in his (1984) *Technical Change and Industrial Transformation: The Theory and an Application to the Semiconductor Industry* Macmillan, Basingstoke, 192, Dosi argues that

> ... technological trajectories are by no means "given by the engineers" alone: we tried to show that they are the final outcome of a complex interaction between some fundamental economic factors (search for new profit opportunities and for new markets, tendencies toward cost-saving and automation, etc.) together with powerful institutional factors (the interests and structure of existing firms, the effects of government bodies, the patterns of social conflict, etc.).

This quotation may, however, be intended to describe only the initial selection of a technological paradigm, rather than the subsequent trajectory which the paradigm, according to a later quotation in the book (ibid., 299), "determines".

Another formulation (Dosi (1982) Technological Paradigms and Technological Trajectories, 154) contains something of the first meaning of "natural", and grants a shaping role for economic factors *after* initial selection of a paradigm, but only within boundaries set by the latter:

> A technological trajectory, i.e. to repeat, the "normal" problem solving activity determined by a paradigm, can be represented by the movement of multi-dimensional trade-offs among the technological variables which the paradigm defines as relevant. Progress can be defined as the improvement of these trade-offs. One could thus imagine the trajectory as a "cylinder" in the multidimensional space defined by these technological and economic variables. (Thus, a technological trajectory is a cluster of possible technological directions whose outer boundaries are defined by the nature of the paradigm itself.)

The usage of "paradigm" is, of course, an extension by analogy to technology of T.S. Kuhn's notion of scientific paradigm. For a useful discussion of the ambiguities of the latter, see Kuhn's postscript to the second edition of *The Structure of Scientific Revolutions* (1970) Chicago University Press, Chicago.

25   Dosi's (1984) *Technical Change and Industrial Transformation*, 68; Dosi, (1982) Technological Paradigms and Technological Trajectories, 153.

26   Thomas P. Hughes (1969), Technological Momentum in History: Hydrogenation in Germany, 1898–1933, *Past and Present*, **44**, 106–32; Bruno Latour (1987), *Science in Action*, Open University Press, Milton Keynes.

27   The closest to a comment on it I have found (again, it may be my ignorance) is this remark in Dosi (1988), The Nature of the Innovative Process, 226:

> To the extent that innovative learning is "local" and specific in the sense that it is paradigm-bound and occurs along particular trajectories, but is shared — with differing competences and degrees of success — by all the economic agents operating on that particular technology, one is likely to observe at the level of whole industries those phenomena of "dynamic increasing returns" and "lock-in" into particular technologies discussed [by Brian Arthur and Paul David].

28   See, for example, Jack Worlton, Some Patterns of Technological Change in High-Performance Computers, *Supercomputing '88*, 312–20.

29   Lincoln interview.

30   This seems, for example, to have been what happened with the ambitious "MP" supercomputer project of Steve Chen of Cray Research, which was cancelled by Cray Research chairman John Rollwagen, because it seemed likely to lead to an unduly expensive machine (Rollwagen interview). Chen left Cray

Research as a consequence, to pursue supercomputing development with funding (but no certainty of the ultimate marketing of a product) from IBM.

31 *An Evolutionary Theory*, 259–61.

32 Paul A. David (1986), "Understanding the Economics of QWERTY: The Necessity of History", in *History and the Modern Economist* (Ed. William Parker), Blackwell, Oxford, 30–49, 43. I am grateful to Peter Swann for the reference.

33 See, for example, W. Brian Arthur (1984), Competing Technologies and Economic Prediction, *Options*, April, 10–13, Arthur writes:

> Very often, technologies show increasing returns to adoption — the more they are adopted the more they are improved. .. When two or more increasing-returns technologies compete for adopters, insignificant "chance" events may give one of the technologies an initial adoption advantage. Then more experience is gained with the technology and so it improves; it is then further adopted, and in turn it further improves. Thus the technology that by "chance" gets off to a good start may eventually "corner the market" of potential adopters, with the other technologies gradually being shut out.

Amongst potential examples discussed by Arthur and David are the qwerty keyboard, the pressurized water reactor, and the petrol-driven, internal combustion, motor car.

34 Barry Barnes (1983), Social Life as Bootstrapped Induction, *Sociology* **17**, 524–45; Barnes (1988), *The Nature of Power*, Polity Press, Cambridge.

35 For some of the reasons why this proved false, see Mel Horwich (1982), *Clipped Wings: The American SST Conflict*, MIT Press, Cambridge, MA.

36 See Christopher Freeman (1988), "Induced Innovation, Diffusion of Innovations and Business Cycles", in *Technology and Social Process* (Ed. Brian Elliott), Edinburgh University Press, Edinburgh, 84–110.

37 Christopher Freeman and Carlota Perez (1988), "Structural Crises of Adjustment, Business Cycles and Investment Behaviour", in *Technical Change*, (Eds Dosi *et al.*), 38–66, 48. These are the general conditions, according to Freeman and Perez, met by the "key factor" of all five successive paradigms they identify, but they clearly believe them to hold for microchip technology as key to the current, emerging, paradigm.

38 Since the first draft of this paper was completed, I discovered an unpublished paper by Professor Arie Rip making essentially the same point about Moore's Law. See Rip (1989), Expectations and Strategic Niche Management in Technological Development (June).

39 Dosi (1984), *Technical Change and Industrial Transformation*, 68. I mention Dosi's study only because it explicitly makes use of the notion of trajectory. Other authors, including for example, Ernest Braun and Stuart Macdonald (1982), *Revolution in Miniature: The History and Impact of Semiconductor Electronics*, Cambridge University Press, Cambridge, make the same assumption (ibid., 103–4, 217).

40 Such a study is beginning at the Technical University Twente, under the supervision of Professor Arie Rip (see note 38).

41   See Cyril Tomkins and Roger Groves (1983), The Everyday Accountant and Researching his Reality, *Accounting, Organizations and Society*, **8**, 361–74. There are of course interesting studies of non-industrial societies to be found within economic anthropology and economic history, though not all by any means delve fully into ethnoaccountancy. See, e.g., Rhoda H. Halperin (1988), *Economies across Cultures: Towards a Comparative Science of the Economy* Macmillan, Basingstoke, and Raymond W. Goldsmith (1987), *Premodern Financial Systems: A Historical Comparative Study* Cambridge University Press, Cambridge.

42   Although it does not deal with technological change, the sociological work that is closest to my argument here is R.J. Anderson, J.A. Hughes and W.W. Sharrock (1989), *Working for Profit: The Social Organisation of Calculation in an Entrepreneurial Firm* Averbury, Aldershot, and Richard Harper (1989), "An Ethnographic Examination of Accountancy" (PhD thesis: University of Manchester). See also Harper (1988), Not Any Old Numbers: An Examination of Practical Reasoning in an Accountancy Environment, *Journal of Interdisciplinary Economics*, **2** 297–306. Another intriguing study is Jean Lave (1986), "The Values of Quantification", in *Power, Action and Belief*, (Ed. John Law), Sociological Review Monograph No. 32 Routledge, London, 88–111. Law himself is currently researching the role of accounting practices in a large scientific laboratory.

43   Aside from the material cited in the previous note, work on accountancy has also been done by sociologists Keith Macdonald and Colwyn Jones: see, e.g., Keith M. Macdonald (1984), Professional Formation: The Case of Scottish Accountants, *British Journal of Sociology*, **35**, 174–89; Colwyn Jones (1989), "What is Social about Accounting?" (Bristol Polytechnic: Occasional Papers in Sociology, Number 7, March).

44   Nelson and Winter (1982), *An Evolutionary Theory*, 411.

45   Alfred D. Chandler, Jr. (1977), *The Visible Hand: The Managerial Revolution in American Business*, Harvard University Press, Cambridge, MA.

46   One reason why this is a dubious way of thinking is that an accountancy system does not come free. A balance has to be struck between the benefits of greater knowledge of one's operations and the costs of such knowledge. It may be that here is a minor replica of the general problem of maximization discussed earlier in the text.

47   Judith A. McGaw (1985), Accounting for Innovation: Technological Change and Business Practice in the Berkshire County Paper Industry, *Technology and Culture*, **26**, 703–25. See also McGaw (1987), *Most Wonderful Machine: Mechanization and Social Change in Berkshire Paper Making, 1801–1885*, Princeton University Press, Princeton, NJ.

48   See, e.g., Anthony F.C. Wallace (1987), *St. Clair: A Nineteenth-Century Coal Town's Experience with a Disaster-Prone Industry*, Knopf, New York. There may be some purchase here on one of the classic debates of economic history, the explanation of the faster rate of mechanization in nineteenth century America compared to Great Britain, for which see H.J. Habakkuk (1962), *American and British Technology in the Nineteenth Century: The Search for Labour-Saving Inventions*, Cambridge University Press, Cambridge. However, it is not clear that British accounting practice was any different from American in the relevant respect. See

Sidney Pollard (1965), *The Genesis of Modern Management: A Study of the Industrial Revolution in Great Britain*, Edward Arnold, London.

49 Robert S. Kaplan (1984), The Evolution of Management Accounting, *The Accounting Review*, **59**, 390–418, 415.

50 Toshiro Hiromoto (1988), Another Hidden Edge — Japanese Management Accounting, *Harvard Business Review*, **66** (4) (Jul./Aug.), 22–6, 22.

51 Karl Marx (1976), *Capital: A Critique of Political Economy*, volume 1 Penguin, Harmondsworth, 932.

52 Freeman (1982), *The Economics of Industrial Innovation*, 167.

53 See, e.g., Latour (1987), *Science in Action*. Space prohibits an exposition of actor-network theory, but one key misunderstanding should be avoided. The actors involved include non-human entities as well as human beings: an actor-network is not a network in the ordinary sociological usage of the term. The concept is closer to philosophical monism than to sociometrics. See the usage of "le réseau" in Denis Diderot (1987), "Le Rêve de d'Alembert", in *Œuvres Complètes* (Ed. D. Diderot), tome 17, Hermann, Paris, 24–209, e.g. 119.

54 It is perhaps significant that such success as neoclassical economics, as enjoyed in the empirical explanation of technological change, seems to be predominantly in the explanation of patterns of diffusion. Some degree of stabilization is a *sine qua non* of the applicability of the concept of diffusion, because there needs to be some sense in which it is the same thing (hybrid corn, or whatever) which is diffusing. For sceptical comments on the concept of diffusion, see Latour (1987), *Science in Action*.

55 There have been calls for some time for the bringing together of the sociology of scientific knowledge and the study of technological innovation, especially by Trevor J. Pinch and Wiebe E. Bijker (1984), The Social Construction of Facts and Artefacts: or How the Sociology of Science and the Sociology of Technology might benefit each other, *Social Studies of Science*, **14**, 339–441. A collection of studies exemplifying the connection has been published — Wiebe E. Bijker, Thomas P. Hughes and Trevor Pinch (eds) (1987), *The Social Construction of Technological Systems: New Directions in the Sociology and History of Technology* Mass.: MIT Press, Cambridge, MA — and a further volume of more recent papers is in press. This body of work has not to date addressed the economic analysis of technological change very directly. The main effort to do so in Bijker, Hughes and Pinch (1987), *The Social Construction*, by Henk van den Belt and Arie Rip ("The Nelson-Winter-Dosi Model and Synthetic Dye Chemistry", 135–58), seems to me insufficiently critical of that model.

56 For "interpretative flexibility", see H.M. Collins (1981), "Stages in the Empirical Programme of Relativism", *Social Studies of Science*, **11**, 3–10. Pinch and Bijker, (1984), The Social Construction of Facts and Artefacts, develop the relevance of the concept for studies of technology, though drawing the more general analogy to "flexibility in how people think of or interpret artefacts [and] in how artefacts are designed" (421).

57 My bias was reinforced by an eloquent presentation of the point by Bruno Latour in an informal seminar at the University of Edinburgh, 6 February 1990.

58   Coombs *et al.* (1987), *Economics and Technological Change*, 6–7.

59   This is the central theme of an old but still valuable paper by Donald Schon, "The Fear of Innovation", as reprinted in Barry Barnes and David Edge (eds) (1982), *Science in Context: Readings in the Sociology of Science*, Open University Press, Milton Keynes, 290–302. The original discussion is of course to be found in Frank H. Knight (1921), *Risk, Uncertainty and Profit*, Houghton Mifflin, Boston.

60   There is no absolute way the distinction can be made *ex ante*. See ibid., 293–4.

61   I have argued elsewhere that there is a productive analogy to be drawn between the testing of technology and scientific experiment as analysed by the sociology of scientific knowledge. See Donald MacKenzie (1989), "From Kwajalein to Armageddon? Testing and the Social Construction of Missile Accuracy", in *The Uses of Experiment: Studies in the Natural Sciences* (Eds David Gooding, Trevor Pinch and Simon Schaffer), Cambridge University Press, Cambridge, 409–35.

62   For such tools, designed for use on non-stabilized networks, see Bruno Latour, Philippe Maugin and Geneviève Teil, "A New Method to Trace the Path of Innovations: The 'Socio-Technical Graph'", unpublished manuscript.

3

EVOLUTIONARY EPISTEMOLOGY AND THE NATURE OF TECHNOLOGY STRATEGY

J.S. Metcalfe and M. Boden

INTRODUCTION

The formulation and implementation of technology strategies by firms in the pursuit of competitive advantage are significant factors determining the nature, rate and direction of much technological advance. These strategy formulation and implementation activities, as practised by the firm, are the focus of this chapter, particularly in relation to the firm's technological development. The success of such an analysis, however, can be seen to depend critically upon a clearer conceptual perception of the nature of technology as it both exists in and is developed by the firm. This chapter, therefore, aims at such conceptual refinement, emphasizing the significance of knowledge, its acquisition, organization and utilization in technological competition. It leads us to the view of strategy formation as a dynamic process by which the firm explores its technological and market environments as a basis for gaining competitive advantage. Much writing and everyday discussion of strategy and policy is devoted to the existence of technological gaps between firms, and whether they are increasing or declining. What is less commonly observed is that the existence of technological gaps is a natural outcome in a world where firms behave differently, and that differential behaviours are the reason why competition is an evolutionary process. Since our context is provided by evolutionary theories of competition, we shall see strategy as an essential aspect of the variety-generating or experimentation process of a firm. Blind variation and selective retention (Campbell, 1987) are the central elements of an evolutionary theory; blind variation does not mean random variation, however. Rather, it entails variation the consequences of which are not entirely predictable *ex ante*. In blind variation in the natural world, variety in the entities is produced without prior knowledge of which ones will have selective advantage, independently of environmental conditions and without learning from past evolutionary successes and failures. Success is as blind as failure. With respect to the development of technology, this notion of blindness needs further and careful elaboration, for variation in the technological sphere is structured and guided by past human experience and by the anticipation of changes in the selection environment. As Vincenti (1990) has emphasized, blindness arises whenever the variation process leads to new knowledge going beyond the limits of foresight or prescience. Several important mechanisms of directed variation are essentially strategic in nature, or so we shall argue in this paper. More precisely, we shall suggest that the understanding of technology strategy can gain much from the application of recent developments in evolutionary epistemology, which provide insights into why firms behave differently and why those differences are limited.

Aspects of the nature of the relationship between technology and the firm having been articulated, attention can be given to the strategy process: the ways in which a firm identifies and selects between a range of technological options; the ways in which these activities are articulated through, and are interrelated with, the firm's technological choices; and the ways they are implemented, and the requisite technological capability accumulated. This process is seen as a cognitive one, with the generation and consideration of options and the choice between them conceivable in terms of the firm forming and testing hypotheses about the most desirable way to continue its technological activities. This issue is explored in some depth, drawing on the concepts of paradigms and research programmes which have been so fruitful in explicating the development of science.

THE MAIN THEMES

Our central organizing theme is that strategy emerges at the intersection of three categories of phenomena: competitive behaviour, organizational design, and technology content and structure. Within the constraints set by these considerations, technology strategy involves the matching of objectives and actions to influence and respond to actual and anticipated changes in the firm's environment. In other words, technology strategy is a relation between the development of technology and the pursuit of competitive advantage in specific organizational and environmental contexts. Although technology is not the only source of competitive advantage for a firm, it is a vital one in the context in which a firm may change drastically its core activities. The ability of the firm to develop its technological position over time, in the context of an evolving envelope of technological opportunity, is widely accepted as a crucial consideration for its long-term growth and survival. Indeed, we see the principal strategic issue as one of balance within the firm: legitimating its current range of activities while creating the set of activities necessary to hold its position in the future.

Within this context our purpose is to locate the discussion of strategy within theories of evolutionary change in epistemology, economics and business: that is, we identify strategy with the generation over time of selective advantages for firms, advantages which are based on the accumulation of specific knowledge capital. In the first part of this paper we explore the triad of competition, technology and organization, making the links with evolutionary theory, drawing upon recent studies in the nature of technology, and leading to the view of strategy as hypothesis formation and experimentation. The second part presents the preliminary results of two case studies on strategy formation. The final part draws together the lessons from these firms in relation to the earlier discussion. A crucial distinction here is between the external selection environment faced by the firm, and the internal selection environment generated by the firm and embodied in its decision rules and communication structures. The maintenance of selective advantage is seen to depend on an appropriate balance between the two domains of evolution. In particular, the internal selection environment determines the range of possible futures a firm considers as options — each a rival hypothesis — and the criteria and process which lead to the choice of one of these intended futures. Moreover, it is the internal environment which conditions perceptions of signals emanating from the external environment and shapes the appropriate response. Correspondingly, the internal selection environment is a principal determinate of the firm's growth and survival.

COMPETITION AND EVOLUTIONARY CHANGE

The starting point for any evolutionary argument is competition, the struggle for relative numerical importance within a market environment. Inherent to the idea of struggle is the existence of differential behaviour. In a world of identical behaviours there is no scope for evolutionary change, and so in textbook economics the only competition which is effectively permitted is that between different industries. The notion of a perfectly competitive industry is in fact a construct which has lost all touch with competition as it is found in the commercial world (Hayek, 1948; McNulty, 1968; Morgenstern, 1972). Central to the theory of evolutionary competition is the dualism between the unit of competitive selection, the firm, and the selection environment. Change can, and indeed must, occur at both levels if evolution is to take place.

In an interesting essay, Toulmin has classified the relevant kinds of change into four categories: calculative, homeostatic, developmental, and populational. The first three of these constitute transformational change, in which it is the units of selection which are subjected to change, while the latter constitutes variational change, in terms of the changing relative importance of the units of selection (Levins and Lewontin, 1983). Calculative change is typified by the theories of rational choice employed within mainstream economic theory, while homeostatic change is change in response to stimuli in accordance with fixed rules. It is equivalent to satisficing behaviour and to the concept of single loop learning (Morgan, 1987). Developmental change is typified by life-cycle theories of organizational behaviour or technology (Abernathy, 1978; Abernathy and Utterback, 1978) and is concerned with the structure and dynamic of development of the selective unit. Finally, population change is typified by "natural" selection, and entails change in the composition or structure of a group of selective units. The crucial point about populational change is that it can only operate in the presence of variety across the competing units, and that variety is generated by changes arising in the first three categories. It is not simply change which is required, but differential change. If all selective units responded in an identical fashion to opportunities and stimuli, the basis for variety would be eliminated.

Within economics, theories of rational calculative choice are dominant. These imply much more than purposive, goal-directed behaviour (Hodgson, 1989). They require that selective units form a comprehensive picture of the relevant opportunity set, translate that opportunity set into a system of preferences which unambiguously ranks every option, and confront that opportunity set with an independently determined choice set. From this confrontation follows the optimal behaviour of the unit. As Shackle (1961) long ago pointed out, choice in this context is purely mechanical, it entails an empty decision. Content to the decision-making process arises from the fundamental facts of imperfect knowledge and the limited interpretive capabilities of individuals. To the extent that choice is rational, it is boundedly rational, constrained by imperfect foresight and the limited computational capabilities of the mind. To the extent that individuals and organizations do optimize, and their purposeful, goal-directed behaviour cannot be denied, their optimizations are local and not global (Winter, 1975; Elster, 1983).

Now the import of this is not counter to rationality, but rather in favour of evolution. For the boundedly rational decisions of individuals and organizations become the basis for differential behaviour. Although they share much of their environment and receive signals in common from that environment, the interpretations based on those signals are rarely the same. Considerable interpretative flexibility is possible — flexibility grounded in the different life histories of individuals and in their different memories. In a recent paper, Wilson (1990) has enriched our understanding of these issues by distinguishing behaviours based on models of reality, in which knowledge equates to representation of an objective world, from those based on what he calls adaptive imaginary representations. In the former class are the classical, rational choice theories leading to global optimization — the Olympian model — together with the more modest schemes based on bounded rationality and local optimization. Clearly the latter are productive of more diverse behaviour, but both seek to relate choice to an accurate perception of objective external circumstances, even if that accuracy is only attained after a period of convergence. A diversity of representations of the world can only be the result of limitations on rational calculative abilities, limitations which constrain the field of vision. By contrast, an adaptive imaginary representation of reality is a set of behavioural rules which greatly simplify reality and provide instructions as to reasonable types of behaviour. These need not constitute an accurate model of reality, but what they must do, if they are to survive, is to motivate adaptive behaviours. In this sense they are judged by their selective value, not by their degree of approximation to reality. Like the genotype, any such set of behavioural rules limits the range of adaptive behaviour which is possible; the adaptive imaginary representation (AIR) is not plastic in the face of environmental change. Indeed, there is no reason at all why a given AIR need reflect changes in the external environment which are, as it were, outside its design range. Diversity in AIRs results in diversity in behaviours in a much richer way than is possible with any global or local model of reality, and from this perspective, ignorance is not inconsistent with adaptive success. This suggests that evolution in the economic sphere is essentially open-ended.

However, evolution is not unbounded, and here the role of organization is paramount. In the field of technological change, creative individuals can rarely achieve their aims acting in isolation; typically they have to operate in the context of organizations if they are to translate ideas into practice, and this is where the other varieties of change become relevant. Organizations operate with rules which limit the permissible kinds of behaviour, and act in homeostatic fashion upon the set of possible options. Organizations also have life histories and accumulated structures and norms of behaviour which self-limit the kinds of change which can be articulated. Organizations are not infinitely malleable or capable of infinitely fine adaptations to opportunities and circumstances. Adaptation is always imperfect with respect to timing and direction, and this is exactly what follows from the facts of imperfect knowledge. There is no more dangerous fiction on which to base our understanding of the economic and social world than that of perfect foresight. Indeed in such a world, creative change is by definition impossible.

Thus the central issues in evolutionary change are the sources of variety on which selection operates, and the limits which operate on the scale and scope of variety. We shall argue that both dimensions are ultimately connected with the strategy formation and implementation process in firms. As we shall suggest subsequently, the central issue here is the process by which the firm constructs its agenda for technological change, and we shall interpret this as a process of guided variation.

Selection environments and the three dimensions of selective performance

In our framework the units of selection are the business activities of firms, defined by sets of products and their associated methods of production. According to the price and performance attributes of these products relative to those of rivals, a business activity will be gaining or losing ground in the market place. Depending on the costs of producing these products the profitability of the business unit is determined, and with it the resource base for expanding capacity relative to rivals and for engaging in other forms of competitive rivalry. Three attributes of the selection environment are thus crucial and relate to the nature of a firm's customers, their sensitivity to differences between rival products, and the frequency with which they make selection decisions (Metcalfe and Gibbons, 1989).

From this perspective the business unit has three attributes: its efficiency and effectiveness, as measured by the characteristics of its products and production methods; its fitness, as measured by its ability to deploy profits to expand production and marketing capacity; and its creativity, its ability to transform its underlying product and process technology.

Now our concern with strategy is predominantly with this latter question of creativity, the ability of the firm to maintain a momentum of technological change relative to its rivals. For it is the firm's strategic vision of where it may go in terms of technology and markets which determines the framework within which a momentum of change can be developed. Of these three dimensions of firm performance, it is obviously the creative one which is of greater long-term significance for the survival and competitive position of the business unit: in the long run, competition is very much a question of the survival of the wisest. But the long run is a cumulation of short run events, and here the roles of efficiency and fitness should not be underrated. Too great an emphasis on creativity at the expense of efficiency may bankrupt the firm, or at least undermine its resource base and hamper its creativity. Too little emphasis on fitness may mean the firm losing market share, and again limit the resource base for future creativity.

Although we focus attention on the business unit as the basic unit of the selection process, it is its products and processes which are the direct subjects of market selection. To paraphrase Dawkins (1986), the artefacts are the replicators, the entities of which copies are made in the production process, and the business unit is the vehicle, the propagating mechanism for the particular technology. Business units are only normally equivalent to firms in the case of smaller enterprises. The modern larger firm is typically composed of an aggregation of business units, in which the centre stands as an umbrella organization, providing some business services and acting as an internal capital market. Firms in this sense are also subject to selective pressure in the broader capital market, through mergers and acquisition. The crucial question then arises of strategy formation at the unit level and strategy formation for the firm as a whole, and the potential conflict which can exist between the two levels.

Before closing this section on evolutionary change, it is important to clarify a number of possible misunderstandings. First and foremost, economic evolution is not the same as biological evolution. The evolutionary analogy is helpful in so far as it emphasizes processes of change which are driven by variety — in our case, behavioural variety with respect to technology. In fact evolution in the economic sphere is as a general rule more rapid than that in the biological sphere, and it permits certain elements, e.g. imitation, which are quite absent in the latter. It is also consistent with a Lamarckian element. Individuals and organizations learn, the organization is a unit for translating the memory of past events and decisions over time; anticipation of future events plays a role in economic decision making quite absent in other evolutionary fields. Far from weakening the evolutionary argument, these considerations enrich it, for they multiply the crucial mechanisms which generate variety. As with other evolutionary spheres, however, the boundary between unit of selection and selection environment remains problematic. Individuals and firms often seek to modify their selection environment in their own favour, e.g. through tariffs, or by influencing government policies with respect to creativity, e.g. through R and D subsidies or collaborative research programmes. Here again, we have the basis for an enriched discussion of evolution.

Technology: context, structure and development

Although technology involves a spectrum of activities and entities (Layton, 1974), we nonetheless find it useful to follow a dualistic approach, distinguishing technology as knowledge from technology as artefact. Our approach to strategy is grounded in this distinction.

Technology as artefact plays a central role in evolutionary theories of competition and industrial change, for it is the products and methods of production of firms which are the primary objects of selection in markets. The artefact dimensions of technology relate directly to the idea of technology as a transformation process in which energy and materials in one form are translated into energy and materials in different forms of a higher economic value (Metcalfe and Reeve, 1990). More precisely, we do not have selection of artefacts, but selection for the performance characteristics embodied in those artefacts. It is the performance characteristics which are valued by users and which convey to the firm a selective advantage, in the same way that the performance characteristics of the firm's production methods convey selective advantage on the cost side. We summarize these artefact dimensions of a firm by the term "revealed technological performance".

Every firm has to pay attention to how the performance dimensions of its artefacts relate to those of its rivals' artefacts. Positioning is a major function of strategy: where the firm is located relative to its rivals, and in which directions to move and how quickly relative to those rivals. In this regard it is convenient to think of the firm's technology as points in a multi-dimensional characteristics space, strategy being concerned with the actions to explore this space. Directions of movement can be defined which give the firm the greatest increment in selective advantage relative to a particular market segment. Clearly this direction must depend on the valuations which customers place on increments in performance, it being possible to have too much of a good thing, as well as too little. Knowing the implicit values of performance characteristics for its main customer groups is a central task of any marketing function in the firm. Notice, though, that the value of performance characteristics for methods of production is often given more explicitly by market data on the price of labour and materials.

It is with respect to performance characteristics that trajectories of technological advance can be identified, in the *ex post* sense of a record of performance at different dates, and, more problematically, in the *ex ante* sense of expected improvements in performance generated by techno-logical forecasts. It is clear that an important role for mapping exercises of this kind is to identify limits on functional performance, beyond which an artefact cannot be pushed. We suggest that mapping of trajectories is a valuable strategic tool for firms.

Our second dimension of technology is technology as knowledge, those concepts, theories, and actions which enable a transformation process to operate. This knowledge is necessarily contained in the minds of individuals who either know directly or know how to find out, e.g. by library search, some piece of information. Here we find the obvious link between technology and the science knowledge base, and the cross-cutting distinctions between different kinds of technological knowledge, such as the codified, tacit distinction versus the procedural, descriptive distinction (Vincenti, 1990).

We can draw a connection here with the idea of technologies as communities of practitioners (Constant, 1984) carrying out their activities within traditions of practice which focus problem solving, suggest solutions, and provide methods for the comparative appraisal of solutions (Laudan, 1984). These traditions reflect the division of labour in the growth of technological capability.

Any firm is typically an embodiment of a number of such traditions, which it seeks to integrate, paying due attention to the boundaries between the conduct of different traditions. Much of this integrating function is summarized by the idea of design activity, and the idea of the technologist as a functional design specialist. One implication of this is that members of the firm belong to wider technological and scientific communities than are contained in the firm. They share educational backgrounds in common, and can communicate readily with other discipline members. However, their knowledge is also focused more precisely by the particular transformation process in which they claim expertise. Specialists in automobile technology may understand little of the details of aeronautical technology, and vice versa, though the broad discipline training of the individuals may be similar. Thus while many elements of the discipline or tradition are codified and public, the specific transformation elements within organizations are extensively tacit and proprietary. This may help explain Von Hippel's (1989) findings on informal knowledge-trading between engineers in the US steel industry, and the general importance of external knowledge networks for many firms.

Important aspects of strategy immediately follow in terms of the discipline/tradition context of the firm's knowledge base, and how this knowledge is to be acquired and accumulated. The issue of external versus internal mechanisms of accumulation is important here, since no firm can expect to command all the knowledge relevant to improving its technology.

Organization

The final element in our triad is the organizational structure of the firm which we see as an operator, translating the skills and knowledge of the individual members into a collective competence. It is this competence which has the dimensions of efficiency, fitness and creativity referred to above, and our concern here is primarily with the creativity-related competence.

The organizational operator is defined in terms of rules of communication, both internally and with respect to the outside world, which filter, transform and store knowledge in the organization. They constitute what Arrow (1978) has called a code of communication, which determines who communicates with whom, about what, with what authority, and with what frequency. As Arrow indicates, there is no single operator which is best in all circumstances, and firms must be expected to vary considerably in their internal codes. Necessarily, the knowledge which is filtered is always a subset of what is available (Jacob, 1982).

Since the operator is highly specific to a given firm, it is not surprising to find that the same individual may perform quite differently in different firms, or that attempts at technology transfer or joint ventures between firms often run into severe difficulties relating to the lack of connectivity of different codes. Moreover, while leadership is always important within firms, it is also operator dependent: the difference between success and failure in entrepreneurship not only depends on the attributes of individuals, but also upon the organizational cooperators with which they work. Conversely, changes in the membership of a firm may entail large variations in its competence.

Once we see the organization in this light, some immediate issues are raised, including: the openness of the organization to external information, and the way in which this is allowed to mould its competence; the capacity of the organization to systematically explore its technological and market environment by entertaining different conjectures about its future; and the ability of the organization to learn from its errors. All of these are well recognized in the literature (Morgan, 1987) and often recur in debates about the relative creativity of large and small organizations (Quinn, 1985) or about the tension between the organizational requirements of efficiency and creativity.

The notion of the organization as competence-defining operator has a further dimension, namely as a competence-changing operator: a competence to change competence (Pelikan). In this respect we shall emphasize its crucial role as an internal selection environment which determines the evolution over time of the firm's knowledge base, and its revealed technological performance. This brings us immediately to the question of technology strategy, drawing upon our triad of concepts.

STRATEGY: PARADIGM, AGENDA AND SELECTION

It is elementary that every business unit must specialize to survive. It cannot do everything, and if it is to be even moderately efficient it must eliminate debate about fundamentals at an early stage in its development. The question then arises of what mechanisms provide the necessary forms for its activities, while permitting the requisite creativity to ensure its long-run growth and survival. More precisely, of what do the fundamentals consist?

In a seminal paper, Dosi (1982) has suggested that technologies have paradigmatic qualities akin to the natural sciences, and defines a technological paradigm as a

> "model" and a "pattern" of solution of *selected* technological problems, based on *selected* principles derived from the natural sciences and on *selected* material technologies.
> (Emphasis in original, p. 152)

Such a paradigm indicates fruitful directions for technological change, defines some ideas of progress, and has a powerful exclusion effect on the collective imaginations of engineers and organizations. A technological paradigm builds cumulatively by suggesting how further progress can be made as each advance in artefact performance opens up new areas of ignorance. Technology progresses by solving a sequence of puzzles in a more or less routine fashion, sometimes guided by theory but often entirely empirically. A progressive technology generates many performance-enhancing puzzles. In this routine aspect it is akin to normal science. Technological design and development is self-evidently more concerned with puzzle solving than with hypothesis testing, more with verifying what works than with theoretical falsification. However, unlike normal science, a technological puzzle is solved when the performance standards of an artefact are improved or become more predictable (Vincenti, 1990), not when the puzzle solution yields a better intellectual understanding of a natural phenomenon. Correspondingly, the criteria for the technological failure of artefacts are more clearly defined in terms of actual or anticipated failure to meet performance standards in a new context or in terms of failure in the economic sense. In science, if Lakatos (1978) is correct, the criteria for failure are rarely clear cut. Similar ideas, referring more or less explicitly to paradigms, can be found in other literatures. Thus Constant (1984) writes of the normal technology of a community of practitioners, a framework which shapes the incremental steps by which a technology is developed. Similarly, Wojick (1979) has used the concept of evaluation policies, that set of decision-making procedures which defines the conceptual framework (administrative, social and economic as well as techno-logical) within which technology is applied to the solution of problems. Wilson's (1990) concept of adaptive imaginary representations is also similar in content, if not in detail.

Each of these approaches is driving at an important point, that the accumulation of technological capability is not random, but structured, by technological and non-technological factors: permitting variety in outcomes while placing limits on permissible variety which can only be transcended by a technological revolution. In all essentials these notions act as discovery and justification heuristics, guiding mental processes and determining questions which are posed (Stein and Lipton, 1989).

However, for our purposes these approaches are not specific enough, either with respect to technology or to the organizational context in which it is articulated. Discussion of strategy requires more precision. It will be useful at this point to distinguish technological design configurations from technological regimes. As we have already suggested, a business unit is organized around some specific trans-formation process which embodies quite specific design principles as to the nature of the product and the method of its production. This interconnected set of principles defines the artefact quite precisely, for instance as a jet engine, a steam turbine, or a petrol internal combustion engine. The set of principles, or design configuration, defines precisely the purpose, mode of operation, construction materials and method of manufacture of the relevant artefacts. It defines the evaluation framework for the technology and it is within the configuration that specific design puzzles emerge. There is considerable evidence that the temporal sequence of innovations which trace the development of a configuration are linked to the patterns of puzzle formation. Sahal's (1981) concept of technological guideposts, Rosenberg's concept of technological bottlenecks (1982) and Hughes's (1985) concept of reverse salients all fit within this idea of the configuration as the puzzle-generating framework. This is more narrowly focused than the broader concept of a paradigm, but the key point is that it is the individual design configurations around which business units specialize. Notice that any design configuration may enable the production of a wide range of products and designs for different market segments, and that it may experience substantial improvements in revealed technological performance over time. A good strategy is one which explores the full potential of a given configuration at least as rapidly as do principal rivals.

From an evolutionary perspective on technological competition, the notion of a design configuration plays a fundamental role, allowing the distinction to be made between competition *within* a design configuration and competition *between* design configurations. How are different design configurations to be distinguished? There are no watertight taxonomic rules to employ in this task. We suggest, however, that different design configurations may produce products aimed at a similar purpose, while basing these products around different but overlapping sets of design principles. Thus the diesel and petrol engines are different design configurations for the internal combustion engine; water jet and air jet looms are different design configurations for weaving cloth, and so on. Typically, different configurations have many elements in common, but differ in crucial technological aspects, resulting in different revealed performance characteristics (Saviotti and Metcalfe, 1984). It is convenient, for descriptive purposes, to draw together different design configurations under the umbrella of a technological regime, and to recognize a further level of competition between different technological regimes based on radically different sets of design principles. Thus, in a sense which we hope is clear, the turbojet is in a different regime from the internal combustion vehicle. The fundamental issue here is the quite different knowledge bases of the technologies in question and the great difficulty firms have in moving between these regimes. At this level we thus have a more precise basis for distinguishing radical from incremental innovations (Freeman, 1985;

Utterback and Kim, 1986) and from changes which are competence enhancing or competence destroying (Tushman and Anderson, 1986).

In a recent paper Henderson and Clark (1990) have taken these concepts further, in terms of a distinction between the component structure of a technology and the architecture of that technology, the way in which the components are brought together to meet the functional need. A design configuration would then be defined in terms of components and architecture and it is their contention that once architectures are agreed on, they become robust and relatively inflexible, while permitting considerable change to take place at component level. In their scheme, radical innovation entails major change at both architecture and component levels, and they are able to distinguish two other categories of innovation: modular (same architecture, different components) and architectural (same components, different architecture). The early history of the tabulating machine industry, a case where technology has strong system elements, illustrates the fruitfulness of these distinctions. The industry developed in the early years of this century around two American companies (subsequently to become IBM and Remington Rand), which were competing intensely with products based on different architectures, one being electro-mechanical and the other mechanical. Within their respective architectures the design configurations were advanced through a long sequence of innovations to a basic set of sub-systems — punch, tabulator and sorter. Up to the early 1930s the performance levels of the two configurations in terms of sorting speed and tabulating speed were similar, but from then on the electromechanical configuration championed by IBM pulled ahead, and by 1938 the company had taken some 80 per cent of the US market. While the sequence of innovations had greatly increased the speed at which punched cards could be sorted, both technologies remained constrained by tabulating speed. It was this bottleneck that the development of the electronic computer relaxed in the later 1940s, even though for twenty years or so computers continued to receive their data input from punched cards. Such is the continuity which shapes even radical change. Perhaps historians do have the best perspective in distinguishing the radical from the incremental (Norbery, 1990; Mokyr, 1990).

In sum, we see design configurations as the fundamental units around which to organize the discussion of technological change. They have Dosi's paradigmatic qualities but they are also equivalent to Constant's (1980, 1984) normal frameworks of technological activity. Gradations of technological change then fall into place, from the incremental change within configurations, to changes in configuration, and most radical of all, changes in regime. These latter correspond to paradigmatic upheaval, especially in Constant's sense of radically recombining basic technological and scientific disciplines, the building blocks of technology, and leading to the creation of a new community of practitioners. They also relate to what Mokyr has termed "macro-innovations", which correspond to the evolutionary idea of punctuated change (Mokyr, 1990).

Underlying much of this discussion are models of the development of knowledge, in particular, Kuhn's concept of paradigm, elaborated in the context of the structure of scientific revolutions, a concept which has provided valuable analogies for the study of technology. He takes a research-based perspective on science, with coherent research traditions springing from the models provided by accepted scientific practice, its theories and applications. While providing an accepted model, a structure for scientific practice, such a paradigm is also open-ended, with a number of unresolved issues. Science thus advances through problem solving within and directed by the paradigm, steadily increasing the scope and precision of the knowledge within it. The existence and discovery of anomaly to the predictions of the paradigm will eventually lead to the destruction of paradigms and their replacement by new paradigms which account for such anomalies and new phenomena. Scientific progress is thus characterized by periods of "normal" science and by exceptional paradigm-redefining revolution. In a similar way we have suggested that much technological advance is in the form of incremental puzzle-solving improvements to a particular design configuration, which will continue until superseded by a radical change in the underlying regime providing a superior substitute, to which many technological practitioners may then adhere to ensure survival.

Of course, Kuhn's paradigm view of normal science is not without its critics (Lakatos and Musgrave, 1970) and some of the central criticisms are pertinent to the concept of normal technology. In particular, the Kuhnian claim that normal science is dominated by a single monopoly paradigm seems out of place in the context of technology. The normal technological situation is for competition between rival design configurations, each with its adherents, for whom it has paradigmatic qualities in terms of beliefs and institutional location. For this reason alone it is worth paying attention to the ideas of Lakatos and his twin concepts of progressive and degenerate research programmes. Lakatos rejects the Kuhnian perspective on scientific revolutions in which there are sudden and irrational changes in vision. Rather, Lakatos maintains, new progressive scientific research programmes come logically to replace their degenerating predecessors. He focuses on particular scientific research programmes consisting of two types of methodological rules: the "negative heuristic", which dictates which research paths are to be avoided, and the "positive heuristic", which determines which paths to follow. The negative heuristic protects an irrefutable "hard core" of the scientific research programme; any anomalies are dealt with in an ordered fashion determined by the positive heuristic, which consists of a partially articulated set of suggestions as to the ways in which the protective belt can be modified. The positive heuristic thus sets out a research programme. The progress of a scientific research programme is essentially continuous, a "proliferation of rival research programmes and progressive and degenerative problem shifts". A research programme may supersede a rival when it explains more. Similarly, on this view, the progress of a strategy is essentially continuous and promotes a sequence of innovations structured within an increasing number of design configurations.

In the case of technology, the basic principles of a technology may also be perceived as constituting a hard core, which is felt to be the best way of meeting a particular need and is thus widely accepted. Practitioners thus work at incremental improvement of this technology, not questioning the basic principles, but working in the protective belt, improving the various features of specific design configurations. The development of an improved technology may well come to displace mature rivals by solving problems in a superior fashion. In accord with Lakatos, the development of technology does not witness sudden irrational changes in practice. Perception and adoption of new technology by communities of practitioners is gradual. Change in practice requires the accumulation of new knowledge, equipment and the concomitant belief in and demonstration of its superiority and potential.

A second difficulty with the paradigm notion as articulated by Dosi is that its organizational context is not specific. We see design configurations as organizationally specific. The business unit has to organize scientific and technological disciplines in an appropriate way, focusing them on generating the required competence with respect to revealed technological performance. This leads us to a different approach to the notion of a paradigm, which relates more directly to the competitive process. This is the idea of a strategic paradigm, through which the firm connects its business objectives with its technological capabilities. It clearly subsumes the technical element, but it must also address broader market objectives, and this can only be done in the context of a specific business unit.

The paradigm is a framework which conditions the internal selection environment of the firm, and firms differ in revealed performance in part because of differences in strategic paradigm. Unlike a scientific paradigm or a technological tradition, a strategic paradigm only requires consensus within the business unit. The broad function of the strategic paradigm is to set out the technological vision of the business unit, that framework which separates desirable from undesirable developments in technology.

Moreover, this paradigm is embodied in the decision-making structures of the organization in three crucial respects: with respect to the determination of the options constituting the technological agenda; with respect to the choice of options from that agenda; and with respect to the model of implementation of the selected options. Thus the strategic paradigm is a framework for non-empty, creative decision. It contains "models" of the business unit's modes of interaction with its external selection environment, and it generates the particular puzzle which drive day-to-day activity. It also generates the examples or standard procedures for problem solving within the strategy.

Thus the strategy paradigm determines the nature of the business unit as an experimental machine, and reflects the decisions of its top competent team (Eliasson, 1988). "What kinds of experiments will be conducted by what methods, over what time frame, and with what resource commitment to internal or external activity?" are the central strategic questions. Experimentation is in this sense a method for exploration of the chosen design configuration. From a social constructionist viewpoint this raises the familiar questions of interpretive flexibility and method of closure within the strategic paradigm (Bijker *et al.*, 1987). It is this which is the crucial role of the strategic paradigm, to act as the generator of hypotheses at the interface of markets and technological capabilities. Thus the strategy paradigm embodies that unique competence concerned with the creativity of the business unit.

It may be helpful to structure this discussion more precisely in terms of models of technology and strategy. One such provisional model, a model which draws on the distinctions made by Remenyi, is presented for illustrative purposes in Fig. 3.1.

In this model, we depict a technological regime as consisting of a hard core of fundamental scientific and engineering principles which are adhered to by any firm working in that regime. They are broad principles which give a coherence to a group of technological activities, and provide the most general statement of the regime. While these principles may be subject to elaboration over time, they are beyond question as agreed principles held in common by all business units operating within the regime. Notice that the hard core consists entirely

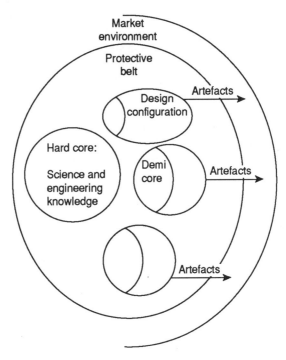

Figure 3.1: A model of technological regimes. (After Remenyi, 1979.)

of knowledge. Centred around the hard core are a number of design configurations, the operational route to designing and producing specific artefacts. These constitute the equivalent to the protective belt. Each design configuration draws on the hard core, but is distinguished from other design configurations by its own specialized hard core or demi-core. Moreover, each design configuration has its own protective belt which provides the freedom to improve existing artefacts and develop new ones. Our central hypothesis has been that business units are specialized with respect to design configurations. The final aspect to note about design configurations is that their protective belt interacts with the external market environment, through the acceptance or rejection of the related artefacts.

Consider now how this model of technology relates to technology strategy. At business unit level, a technology strategy builds from the acceptance of the demi-core of a design configuration — a set of principles which provide the positive and negative heuristics concerning which artefacts will be produced and developed. No artefacts can be entertained which do not fit within the demi-core of the configuration, so the latter acts as an exclusion principle for patterns of technological development. Commonly, the demi-core also guides the patterns of incremental technological change which are legitimate. Thus, the demi-core on the one hand gives the benefits of a focus to creative activity, while at the same time partially blinding the business unit even to developments elsewhere in the regime's protective belt, let alone the threat of competition from different regimes. At the level of the firm proper we may have a collection of design configurations which build from the strategic hard core of the firm. Hence, with this model of strategy, there is a strong element of self similarity between the levels of firm and business unit.

Some familiar strategic questions then fall into place. How should R and D activity be divided between the demi-core and the protective belt at the level of design configurations? Should demi-core work (essentially elaborations) be located on a central R and D facility, or in the operating business units? Similarly at the firm level the same issues arise, in terms of the remit of a central R and D department. For the firm which is operating with several design configurations, the issue arises of how synergy is developed with respect to the strategic hard core, and how best practice is diffused across different business units. For such a firm, incremental change of technology takes place within design configurations, while radical change entails the addition of new design configurations, and a new demi-core. Because of the differences in the content and structure of different demi-cores, radical innovations are almost invariably handled more effectively in the context of a separate, additional business unit.

While the above is only one model of the relation between technology and technology strategy, it has the merit of placing a number of questions within the same framework. Most important of all, it helps understand stability and change with respect to a firm's technological activities, and thus the limits of technology strategy. More than that, it is suggestive of the compelling power of the logic of technological regimes and configurations in determining how and why firms behave differently.

Differences in the long-run competitive performance of firms, we argue, are inextricably related with their strategic paradigms. These set the vision and lead to the choice of the design configurations from which all else follows. Unless the internal selection mechanisms generate hypotheses and experiments consistent with the external selection environment, the business unit cannot expect to enjoy selective advantage.

At this point the question naturally arises of what happens when the strategic paradigm fails, either in a functional sense, with increasing difficulty of improving performance, or in a presumptive sense, with the emergence of new scientific evidence or superior design configurations operated by competitors. Our study has shown how the strategy of dividing commitments between an existing and a threatening design configuration is often followed, but is rarely successful. Neither strategic paradigms, organizational operators nor knowledge bases appear to be sufficiently plastic for this to succeed (Cooper and Schendel, 1976). Be that as it may, we suggest that the overthrow of a strategic paradigm requires the development of different organizational structures, and that this is the test for such a revolution. Either a new business unit is established by an established firm, or the existing activities are radically restructured, or a group of new entrant business units emerges, each of which are the *sine qua non* of a strategic paradigm change.

In closing this chapter it is worth returning to our opening themes of evolutionary change and strategy as the interaction between competition, technology and organization. Although evolutionary change is driven by technological variety, it is the creative strategies of firms which stimulate that process. We have suggested that strategies have a localized paradigmatic quality within each organization and that they are embodied in the rules and structures which define the organization's translation of skills into competence. This paradigm generates variety, but it also constrains the kinds of varieties which can be entertained. Normal technology generates hypotheses which are essentially conservative and possess inertia in the face of change.

Note

Our ideas in this chapter derive from an ongoing research project on technology strategy and competitive advantage in British industry. As well as conceptual refinement of the issues described above, the project has an empirical component which aims to provide evidence to support these ideas, through examination of the ways in which a selection of firms identify their technology choice sets, the choice procedures and choice criteria used and the ways in which technological capability is accumulated. It consists of a series of case studies of major UK firms wherein international competitive performance is seen to depend on sustaining a momentum of technological change. Consequently, the firms under investigation — ICI, Lucas Industries, Rolls-Royce and Pilkingtons — are large R and D spenders with diversified and science-dependent portfolios.

We gratefully acknowledge financial support from the ESRC programmes on New Technology and the Firm to the conduct of the research reported here.

References

Abernathy, W. (1978) *The Productivity Dilemma*, Johns Hopkins University Press, Baltimore, MD.
Abernathy, W. and Utterback, J.M. (1978) Patterns of Innovation in Technology, *Technology Review*, **80**.
Arrow, K.J. (1978) *The Limits of Organization*, Norton.
Bijker, W.E., Hughes, T.P. and Pinch, T. (1987) *The Social Construction of Technological Systems*, MIT Press, Cambridge, MA.
Campbell, D.T. (1987) "Blind Variation and Selective Retention, in Creative Thought as in other Knowledge Processes", in *Evolutionary Epistemology, Theory of Rationality and the Sociology of Knowledge* (Eds G. Radnitzky and W. Bartley), Open Court.
Constant, E.W. (1980) *The Origins of the Turbojet Revolution*, Johns Hopkins University Press, Baltimore, MD.
Constant, E.W. (1984) "Communities and Hierarchies: Structure in the Practice of Sciences and Technology", in *The Nature of Technological Knowledge: Are Models of Scientific Change Relevant* (Ed. R. Laudan), Reichter, Dordrecht.
Cooper, A.C. and Schendal, D. (1976) Strategic Response to Competitive Threats', *Business Horizons*, 61–9.
Dawkins, R. (1986) *The Blind Watchmaker*, Longman, Harlow.
Dosi, G. (1982) Technological Paradigms and Technological Trajectories, *Research Policy*, **11**.
Eliasson, G. (1988) "The Firm as a Competent Team", mimeo, IIESR, Stockholm.
Elster, J. (1983) *Explaining Technical Change*, Cambridge University Press, Cambridge.
Freeman, C. (1985) *Technology Policy and Economic Performance*, Pinter, London.
Hayek, F. (1948) "The Meaning of Competition", in *Individualism and Economic Order*, Chicago University Press, Chicago.
Henderson, R.M. and Clark, K.B. (1990) Architectural Innovation: The Reconfiguration of Existing Product Technologies and the Failure of Established Firms, *Administrative Science Quarterly*, **35**.
Hodgson, G. (1989) "Evolution and Intention in Economic Theory", in *Evolutionary Theories of Economic and Technological Change: Present Position and Future Prospects*, (Eds P. Saviotti and S. Metcalfe) (1991) Harwood, London.
Hughes, T.P. (1985) *Networks of Power*, John Hopkins University Press, Baltimore, MD.
Itami (1987) *Mobilizing Invisible Assets*, Harvard University Press, Cambridge, MA.
Jacob, F. (1982) *The Possible and the Actual*, University of Washington Press, Washington DC.
Kuhn, T.S. (1962) *The Structure of Scientific Revolutions*, Chicago University Press, Chicago.
Kuhn, T.S. (1977) *The Essential Tension*, Chicago University Press, Chicago.
Lakatos, I. (1978), *The Methodology of Scientific Research Programmes*, Cambridge University Press, Cambridge.

Lakatos, I. and Musgrave, A. (Eds) (1970) *Criticism and the Growth of Knowledge*, Cambridge University Press, Cambridge.

Laudan, R. (1984) "Cognitive Change in Science and Technology", in R. Laudan *op. cit.* (Constant, 1984).

Layton, E. (1974) Technology as Knowledge, *Technology and Culture*, **15**.

Levins, R. and Lewontin, R. (1983) *The Dialectical Biologist*, Harvard University Press, Cambridge, MA.

Metcalfe, J.S. and Gibbons, M. (1989) "Technology, Variety and Organization", in *Research in Technological Innovation, Management and Policy*, (Eds R. Rosenbloom and R. Burgleman), Vol 4, JAI Press.

Metcalfe, J.S. and Reeve, N. (1990) "On Technology Taxonomy", mimeo, PREST, Manchester.

Mokyr, J. (1990) *The Lever of Riches*, Oxford University Press, Oxford.

McNulty, P. (1968) Economic Theory and the Meaning of Competition, *Quarterly Journal of Economics*, **82**.

Morgan, B. (1987) *Images of Organization*, Sage, London.

Morgenstern, O. (1972) Thirteen Critical Points in Contemporary Economic Theory: An Interpretation, *Journal of Economic Literature*, **4**.

Norbery, A.L. (1990) High Technology Calculation in the Early 20th Century: Punched Card Machinery in Business and Government, *Technology and Culture*, **32**.

Pelikan, P. (1989) Evolution, Economic, Competence and the Market for Corporate Control, *Journal of Economic Behaviour and Organization*, **12**.

Quinn, J.B. (1985) Managing Innovation: Controlled Chaos, *Harvard Business Review*, May/June.

Remenyi, J.V. (1979) Core demi-core interaction: toward a general theory of disiplinary and subdisciplinary growth, *History of Political Economy*, **11** (1).

Rosenberg, N. (1982) *Inside the Black Box*, Cambridge University Press, Cambridge.

Sahal, D. (1981) *Patterns of Technological Innovation*, John Wiley, Chichester.

Saviotti, P. and Metcalfe, J.S. (1984) A Theoretical Approach to the Construction of Technological Output Indicators, *Research Policy*, **13**, 141–51.

Shackle, G.L. (1961) *Decision Order and Time in Human Affairs*, Cambridge University Press, Cambridge.

Stein, E. and Lipton, P. (1989) Where Guesses Come From: Evolutionary Epistemology and the Anomaly of Guided Variation, *Biology and Philosophy*, **4**.

Toulmin, S. (1981) "Human Adaptation", in *The Philosophy of Evolution*, (Eds U. Jensen and R. Harre), Harvester Press, Brighton.

Tushman, M. and Anderson, L. (1986) Technological Discontinuities and Organization Environments, *Administrative Science Quarterly*, **31**.

Utterback, J.M. and Kim, L. (1986) "Invasion of a Stable Business by Radical Innovation", in *The Management of Productivity and Technology in Manufacturing*, (Ed. P. Kleindorfor), Plenum Press.

Von Hippel, E. (1989) *Sources of Innovation*, MIT Press, Cambridge, MA.

Vincenti, W.G. (1984) Technological Knowledge without Science: The Innovation of Flush Riveting in American Airplanes, *Technology and Culture*, **25**.

Vincenti, W.G. (1990) *What Engineers Know and How They Know It*, Johns Hopkins University Press, Baltimore, MD.

Wilson, D.S. (1990) Species of Thought: A Comment on Evolutionary Epistemology, *Biology and Philosophy*, **5**.

Winter, S. (1975) "Optimization and Evolution in the Theory of the Firm", in *Adaptive Economic Models* (Eds R. Day and E. Groves), Wiley, Chichester.

Wojick, D. (1979) "The Structure of Technological Revolutions", in *The History and Philosophy of Technology* (Eds G. Bugliarello and D.B. Doner), University of Illinois Press.

4

THE DYNAMICS OF TECHNO-ECONOMIC NETWORKS

Michel Callon

INTRODUCTION

Over the past ten years, sociologists and economists have reached a similar conclusion by different paths: scientific and technical creation, as well as the diffusion and consolidation of its results, stem from numerous interactions between diverse actors (researchers, techno-logists, engineers, users, industrialists).[1] The problem then arises of analysing these interactions and accounting for the choices made. How can we explain the fact that in certain cases, trajectories are successful and stabilize, whereas in others new configurations appear? As yet there are no satisfactory answers to these questions. In order to develop a deeper analysis, we propose using the concept of the techno-economic network, whose study will cast a new light on irreversibility and will explain the particular role played by technology.

The analysis of science and technology lies at the heart of the debate about irreversibility, or perhaps what should be called the processes of irreversibilization and reversibilization. On the other hand, technology constitutes one of the main sources of constraints: it creates systems (Gille, 1978; Hughes, 1983), it produces network externalities (Katz and Shapiro, 1986) and, through its localizing effects, it permanently closes off certain options (Arthur, 1989). Studies of the economics of techno-logical change have shown that a number of the hypotheses of the standard model, including reversibility, are unable to account for the observed phenomena (Foray, 1989). But technology and science are also causes of radical change and uncertainty (Bijker et al., 1987), generating irreversibilities and constituting a powerful tool for creating reversibility by producing many new options (MacKenzie and Wajcman, 1985). In this chapter, we would like to show that it is possible by the same means to measure both these mechanisms of irreversibilization and reversibilization. Further, this measurement will enable us to give a new formulation of the passage from the micro to the macro.

In this introduction we will make do with a provisional definition: a techno-economic network (TEN) is a coordinated set of heterogeneous actors — for instance, public laboratories, centres for technical research, companies, financial organizations, users and the government — who participate collectively in the conception, development, production and distribution or diffusion of procedures for producing goods and services, some of which give rise to market transactions. In certain cases, it is possible to anticipate the evolution of these TENs — the actors behave predictably, and the technology and its products evolve along lines that are relatively easy to characterize. In other cases, however, the actors composing TENs have significant degrees of freedom. They develop complicated strategies, there may be a number of innovations, and these provoke unexpected rearrangements. They can separate into smaller networks, or they can join other TENs to form more or less extensive ones. How can we account for the emergence, increase, closure and dismemberment of TENs?

We shall first present the analytical tools that will enable us to understand and describe the mechanisms by which heterogeneous activities are brought into relationship with one another. We will introduce the concepts of intermediaries, actors and translation. We shall then show how networks are established and evolve. Here the two central concepts will be convergence, which deals with the construction of a unified space for elements that are in principle incommensurable, and irreversibilization, which enables us to consider the longevity of these connections and the predetermination of their evolution. Finally, we shall look at the dynamics of TENs. We will analyse the diversity of possible trajectories, at the same time underscoring the fact that the definition of the actors themselves (their identity, their skills) is closely linked to the state of the network, and that the same holds for the (qualitative or quantitative) tools used to describe them. We will then suggest the operation of a fundamental clustering mechanism within networks.[2]

FROM INTERMEDIARIES TO ACTORS

Techno-economic networks are organized around three poles:

1 The scientific pole which produces empirical knowledge. This pole consists of universities and other independent research centres (public and private). Industrial research laboratories belong to this pole, to the extent that their activities are similar to those of university research centres.

2   The technical pole, which conceives of, develops or transforms artefacts destined to serve specific purposes. Examples of these outputs are models, pilot projects, prototypes, tests and trials, patents, and standards. Members of this pole include technical laboratories in companies, cooperative research centres, and pilot plants.

3   The market pole, which contains users who more or less explicitly express (produce) a demand or needs, and try to satisfy them.

The processes of production and exchange that we can observe taking place in TENs involve a whole series of activities of intermediation between these poles.[3] The ultimate incorporation of science into technology involves *transfer operations*. The mobilization of technology to satisfy potential or expressed demands in the market takes the form of activities which by convention we will call *development/distribution*. In general, these are carried out by companies and their marketing networks.

The different poles have memberships, goals and procedures which may apparently be mutually exclusive. For instance, a researcher working on the fine structure of ceramics, and a user looking for a comfortable, fuel efficient, reliable car with good acceleration might seem to belong to completely different worlds. In practice, however, arrangements and links are made between the members of different poles, so that the outputs of various activities are exchanged with the members of other poles. If we want to understand how these activities are brought into relation with each other, we have to explain the creation of a common, unified space between these heterogeneous poles. To do so, we will have to draw on sociological theory and on economics.

In economics it is *things* which bring actors into relationship with each other. A producer and a consumer enter into a relationship via a product that one supplies and the other demands. This situation can be generalized by means of the notion of an intermediary. An intermediary is anything which passes from one actor to another, and which constitutes the form and the substance of the relation set up between them — scientific articles, software, technological artefacts, instruments, contracts, money, etc.[4]

In sociology, the behaviour of actors is intelligible only within the context in which they are being considered: the actor and the system (Crozier and Friedberg, 1977); historicity (Touraine, 1974); rules (Reynaud, 1989); the agent and the field (Bourdieu, 1980); roles and functional needs (Parsons, 1977); etc. No actor can be dissociated from the relationships that actor enters into.

By drawing together the viewpoints of economists and of sociologists, we have actors who recognize themselves in interaction. This interaction is embodied in the intermediaries that the actors themselves put into circulation.[5]

Intermediaries as networks

For our purposes, the range of intermediaries can be classified into four types:

1 Texts, such as reports, books, articles, patents, notes etc. These are material goods, requiring media (paper, disks, magnetic tapes, etc.) which can withstand transport and assure a degree of immutability.[6] Here we will be particularly concerned with scientific texts.
2 Technical artefacts (scientific instruments, machines, robots, consumer goods, etc.) which are (relatively) stable, organized groups of non-human entities which cooperate in the fulfilment of certain functions, carrying out certain tasks.
3 Human beings and the skills (knowledge, know-how etc.) they incorporate.
4 Money, in all its different forms.

We will show that each intermediary, from whatever category, describes (in the literary sense of the term) and composes (in the sense of giving form to) a network of which it in a way forms the medium and to which it gives order.

Texts

Let us now consider texts, and, in particular, scientific texts (Callon *et al.*, 1986; Latour, 1989). A scientific text establishes branches and connections to all other sorts of texts and literary inscriptions. Audiences are identified by the choice of journal, of language, of title. Even very simple inscriptions like journal name, title, author names, can give indications about collaboration or relative contributions to particular programmes. Citations show the links of the work described in a particular text to a series of other texts. Words, ideas, concepts, either already known or completely new, define each other, in the course of the text. The text might portray electrons, enzymes, government agencies, bizarre oxides, procedures for synthesis, experimental set-ups, powerful companies like IBM, entire industrial sectors. All these actors intermingle and are transformed into linked destinies, "socio-technical dramas". The text may refer back to other texts associating the actors in different forms, and extending the initial network.

This notion of a text is essentially different from that of one closed on itself, subject to the classic opposition between context and content. The text is seen as an object that defines and associates heterogeneous entities, their performance and their skills: the scientific text is itself a network, whose description it furnishes.[7]

This equivalence between a text and the network it describes, which has been meticulously established by sociologists of science, can be extended without difficulty to cover all the various inscriptions that circulate along TENs, from diagrams and working notes within laboratories, to patents,[8] user manuals, catalogues, marketing studies, and so on. We have concentrated on scientific texts because they have played an ever more important role in the networks we are interested in. Indeed, economic activity could almost be described as the activity of producing marketable goods from the basis of scientific texts! It could also be concluded that when we make an intermediary equivalent to a network, all we are doing is highlighting the descriptions that it provides, creating a suitable context.

Technical objects

Various non-human entities (machine tools, internal combustion engines, video machines, nuclear plants, train tickets distributors, etc.) can be described in terms of networks linking heterogeneous actors. Recent developments in the sociology of technology, and in particular pioneer work by Madeleine Akrich and Bruno Latour, enable us to understand how this happens. A technical object can be assimilated into a programme of action coordinating a set of complemenary roles fulfilled by non-humans (who constitute the objects) and humans (producers, users, repairers, etc.) or other non-humans (accessories, integrated systems) which form its peripherals or extensions. It is not hard to reconstruct the kind of descriptions that these programmes evoke. All one has to do is look at the object as it is being used, identifying the various organs or actors who intervene, determining what they are doing and the way in which they communicate, give each other orders, interrupt each other, observe certain protocols. These descriptions — or "textualizations" — of networks coordinated by technical objects are not so infrequent as one might believe. Textualization, which in a sense empowers a group of non-humans with speech, may be seen to operate quite frequently. We will look at two instances.

Clear examples of the role of non-humans in networks can be found in the phases of development and disputation (Akrich, 1987; Callon, 1981; Latour and Coutouzis, 1986; Law, 1988; Law and Callon, 1988). When an object is still at the project stage, it is continually under discussion: what should its characteristics be? What should it be used for? What should it do? What skills should its users have? Who should intervene in its maintenance? These debates are always socio-technical. As has been shown elsewhere, engineers can transform themselves into sociologists, historians, moralists or political scientists at the very moment when they are caught up in the most technical of design tasks. For instance, should a car be considered simply as an economical means of transport, without frills, or is one of its basic features the satisfaction of repressed desires (Callon, 1987)? Is it reasonable to tolerate user intervention when a solar battery lighting kit breaks down, or should it be hermetically sealed to prevent the danger of amateur do-it-yourselfers wrecking it (Akrich *et al.*, 1987)? When responding to each of these questions, designers are making decisions which are inseparably both technical and social in nature. While it is being defined, the technical object is continually being reinserted into various socio-economic contexts, which constitute different possible network configurations.

The network ascribed to the object is set out and inspected in a number of situations. The roles played by the human actors with respect to the artefacts, and the connections made with other technical objects are revealed. Thus the machine is interpreted and deconstructed — that is to say put into context.[9]

The interaction and coordination of non-humans and humans in networks is recorded in, for instance, maintenance manuals, codes of procedure, and other user manuals[10] (Akrich, 1989b). In some cases operation of machines requires specialist skills, and the use of different coloured signals, or texts in the form of written legends on the machine itself. These machine operations require human interaction, as well as emotional and even moral reflexes (Latour, 1988). An artefact is never the enigmatic and distant concoction it is too often reduced to. When it comes into contact with its user, it carries on a stream of discourse, and it displays the scars of the various textualizations that have accompanied its design and displacement (Akrich, 1989a).

This capacity of technical objects to distribute roles to humans and non-humans (in a more or less constraining and explicit way) and to link them together (that is to say, to create a network) means that they can be assimilated within programmes for action which necessarily produce literary records, even if these may take various different forms. Here again, the network can be read in the object.

Skill

Incorporated skills can be of many different types, ranging from the ability to mobilize at any time a network of social relationships (that is to say humans), to purely technical skills, whose essential quality is to be able to control artefacts which, without these skills, would not function (for instance, programmers expert in computer systems; skilled workers who have sufficiently disciplined themselves that they can work without any great risk in a long chain of automated machinery). In other words, it is impossible to adequately describe a skill without reconstituting the relevant network(s) made up of humans, texts and machines (Cambrosio and Limoges, 1990; Mustar, 1989). Once again, the act of description reveals the contexts.

Money

The traditional functions of money include being an instrument of exchange and a value reserve. In the former case, there is necessarily some reciprocal return from the recipient of the money to the giver. Money is the basis for a minimal and essential return of information. It stabilizes and sanctions the relationship between provider and client that other categories of intermediary have proposed. The discipline of economics is constructed on the systematic analysis of this relationship, and on what it says about the actors involved and their commitment to each other. Of course, information is not the only form of return; recognition, reputation, legitimacy, allegiance and belief are other possible returns. However, we will not examine them here, since they play a secondary role in the networks we are dealing with.

With respect to the reserve value of money, things are even clearer. Money allows private and public funding of projects; when it does so, the flow of money is accompanied by orders, indications and recommendations which bring out, define and link a whole series of human and non-human actors (Aglietta and Orlean, 1982). The following constitutes an example of a statement accompanying a flow of money: cooperate with X at Thompson and Y from laboratory Z to achieve a critical temperature of 150°K, and you will get a loan of $A. Again, money links a set of roles, acting as a network itself.

Hybrid intermediaries

The categories of intermediary described so far may be termed pure. In real life, however, one normally finds *hybrid* intermediaries. For example, the process of textualization is generalized and we enter into a "civilization" of inscriptions that covers all forms of intermediaries. The more one writes, the more one links, and the more one links, the more one writes. Similarly, the hybridization between human and non-human actors can become so intense that one can hardly distinguish between these two types of intermediary. The best example of this type of hybridization is provided by systems of distributed intelligence, which mixes together computers requiring programmers, and programmers mobilizing computers.

Impurity is the rule, and is exemplified particularly well in the service sector. In this sector humans and non-humans form groups of intermediaries that are exchanged together as packages. For example, a package holiday involves many computers, alloys capable of supporting the thrust of jets at the moment of take-off, research departments, market studies, advertisements, welcoming air stewards and stewardesses, locals who smile as they carry visitors' luggage, bank loans, etc. The analysis of intermediaries that we are proposing enables us to study the economics of both "material" goods and "non-material" services with the same theoretical equipment. All that is required is that one accepts the addition of a few extra texts and personnel to the groups in question.

Decoding intermediaries
The previous considerations show that each intermediary (whether pure or hybrid) more or less explicitly and consensually describes a network, that is to say a collection of human and non-human, individual and collective entities (defined by function and identity) and the relationships into which they enter. This has two consequences. The first concerns the crucial role played by intermediaries in the establishment of the social link, to which they give existence and consistency: the actors define each other by way of the intermediaries they put into circulation. The second is methodological: the social link can be read in the inscriptions which mark the intermediaries. During the Renaissance, much delight was taken in reading the great Book of Nature. We should extend the metaphor, and feverishly try to "read" all the intermediaries that pass through our hands — artefacts, scientific texts, organizations, and cold money. In this context sociology is simply an extension of the science of inscriptions; it should turn its attention partly away from the actors, to look at the intermediaries and consider what makes them act and speak as they do.

Actors

By "actor" we mean any entity able to associate the various elements that we have listed so far, and which defines and constructs (more or less successfully) a world peopled with other entities, gives them a history and an identity, and qualifies the relationships between them. If we held to this definition alone, it would not be wrong to say that an intermediary can be an actor. Why, then, is it necessary to reintroduce the notion of an actor? Why not make do with that of the intermediary? We will see that this distinction is essential as soon as we try to account for attribution mechanisms.

Any interaction includes a mechanism for the attribution of intermediaries. Further, this attribution is often inscribed in the intermediaries themselves. The scientific article, for instance, is signed, and the technical object is trademarked. Incorporated skill is attributed, under our law at least, to the body itself and to the subject who is said to "animate" it. One of the essential elements of the description contained in an intermediary is the identification of the actor who claims attribution of the author's rights. This identification, like all the other hypotheses made by the intermediary, is always open to dispute and question. It is no less controversial than the other elements in the network. Its solidity or legitimacy depend on the conventions that it presupposes, and without which the potential change would be improbable. It is impossible to make an absolute distinction between an actor and an intermediary, except for the attribution mechanisms attached to the former: an actor is an intermediary attributed with putting other intermediaries into circulation.[11] Thus, an actor can be described as a transformer, producing (by combination, mixture, concatenation, degradation, computation, anticipation, etc.) an N+1st generation of intermediaries out of generation N. Researchers transform texts, experimental apparatus, and grants into new texts; companies combine machines and incorporated skills so as to give rise to products or services which are put into circulation, for use by consumers, who play certain roles.

Is a group an actor or an intermediary? Is an actor a force for conservation or for transformation? Distinguishing between actors and intermediaries has nothing to do with metaphysics, ontology, or the philosophy of the rights of man. It is above all an empirical problem whose solution is to be found in observation.

Actors, just like intermediaries, can be hybrid, combining different elements. They may be collective or otherwise. In any case, the observer is constrained to hypothesize an ontology with variable content and geometry. This variability on the part of the actor applies to all forms of groups. It holds equally for companies and for associations between humans and non-humans.

NETWORKS OF TRANSLATION

Any group, whether actor or intermediary, describes a network, that is to say it identifies and defines other groups, actors and intermediaries, as well as the nature and form of the relationships that unite them. These descriptions constitute so many stopping points, asymmetries or folds (Deleuze, 1989). An actor A transforms intermediaries of rank N with which it agrees to be involved, into intermediaries of rank N+1. The modified intermediaries are so many scenarios, bearing their author's signature, looking for actors ready to play the roles they inscribe. What emerges then is an actor-network, which can be written in the form $N(A) = Br^1C, Cr^2E, Fr^3H, Kr^4L$. N(A) signifies that we are dealing with a network which A is attributed with putting into circulation. We can say of A that it defines (in the form of intermediaries interposed) a series of entities/groups B, C, E, F, H, K, L (actors or intermediaries, humans or non-humans, pure or hybrid, collective or not) and at the same time the relations, r^1, r^2, r^3 and r^4 that link them. N(A) is nothing other than the action itself that constructs networks (either consolidating networks already in place, or developing new ones) by putting intermediaries into circulation. We have no need of any further definitions.

Once we have established the distinction between actors and intermediaries, and also their close interdependence, we still have to resolve one problem. How do the different actor-networks, which have no a priori reason to be compatible with one another, reach agreement? Why do at least some of these agreements have a considerable durability? It would be possible for B not to accept the definition of itself implied by a relation with A, or for C to attribute another identity to B. The answer to this question is to be found in the process of the convergence and irreversibilization of techno-economic networks. Before presenting this, we need to develop an analysis of the elementary relationship established between two actors A and B, which we call the translation operation (Callon, 1976, 1980, 1986, 1989; Callon and Law, 1982; Latour, 1984, 1989; Law, 1986).

Translation

The operation of translation is performed by an entity A on another entity B. Both A and B can be actors or intermediaries, humans or non-humans. The statement "A translates B" can have two different meanings. First of all, it means that A provides a definition of B. In so doing, A may impute B with certain interests, projects, desires, strategies, reflexes or afterthoughts. A chooses amongst all these possibilities, but this does not mean that A has total freedom. What A does or proposes is consequent to a whole series of intertwining translation operations, some of which determine ensuing translations to the point of pre-programming them. The general rule is that one actor translates several others, between whom it establishes relations. Coming back to the preceding notation, we could say that A translates B, C, D and E. These latter are in some sense interdefined by A, since from all the evidence, what B is depends on its relations with C and D. At the same time as it defines B, C, and so on, A also defines itself.

These definitions, and this is the second dimension of translation, are always inscribed in intermediaries. This follows directly from the preceding analysis. These intermediaries can equally well be round-table discussions, public declarations, texts, technical objects, incorporated skills, or money. It does not make any sense to speak of translation in general: one has to start by defining the medium, the material into which it is inscribed. A translates B. A might be the company that has conceived of, produced and distributed a machine, and B the users, satisfied or not, who occupy the positions thus created for them as users. Alternatively A may be the author of a scientific paper, of which B might be, for example, the target audience, or the enzyme whose performance is described. In another case A may be the lender of a bank loan to B. Clearly translation involves three terms: A → I (intermediary) → B.[12]

Translations change over time. Sometimes they succeed in establishing a compromise, which is the fruit of iterations and more or less difficult and long negotiations (Akrich et al., 1987). This compromise matches the definition of B by A with that of A by B, inscribing them in intermediaries (texts, machines, incorporated skills etc.). These latter become their support, their more or less faithful executive. The detour that translation operates — we are using this term in the sense that economists speak of production detours — can be more or less complicated. It can involve anything from the isolated, homogeneous intermediary up to the hybrid intermediary consisting of a cascade of intermediaries interposing a whole series of articulated roles between A and B, each role being linked to the other by intricate hooks and feedback loops. Translation places the interdefinition of the actors and its inscription in intermediaries at the heart of the analysis. It extends traditional definitions of action.

The process of irreversibilization

One advantage of the preceding definitions is that they do not establish any single solution to the problem of providing continuity between the network and the actor. When two translations link together, they form a third translation. A translates B, C and D; B in turn translates C, E, F and M; C translates E, G, M; D translates C, F, G, Q . . . . And, of course, there is no need to limit oneself just to A's point of view or to make it the principle of the network's organization. A might be translated by X, Y and Z, and retranslated by F or C. A network can be formed by the aggregation of all these generally polycentric actor-networks, in which intermediaries circulate and link each other. They belong to the different categories we have distinguished, and propose ranges of more or less compatible, more or less contradictory translations. Thus behind the heterogeneity of the actors we can find textualizations and descriptions which sometimes agree with each other, and form linkages. It is in these processes that commensurability, if it exists, has to be sought — not in the actors' cognitive capacities.

We have now progressed sufficiently to commence the description of these complicated dynamics. Two concepts will be useful to us for producing this description: those of convergence and irreversibility. Both are involved in the acts of translation and the networks that they sometimes succeed in forming.

Convergence

The concept of convergence concerns the degree of accord engendered by a series of translations, and by the intermediaries of all sorts who operate them. At the same time it reveals the frontiers of a TEN. It thus operates in two respects: alignment and coordination.

In order to define alignment, we will look at the elementary translation operation A→I→B. The intermediary I and the definitions of A and B that it offers will find more or less wide acceptance or be more or less contested. In situations of controversy or conflict there is the saying "traduttore-traditore" — translation is betrayal. Translation is sometimes denounced in this way — workers do not want to fulfil the role assigned them by the machine, users deprecate the quality and use of a given product or service being offered them, scientists counter their fellow-scientists' arguments, electrons no longer pass from one electrode to another. Symmetrically, the inventor denies his or her innovation, the author exclaims: "I only spoke about the memory of water in order to arouse your curiosity, to express my flight of fancy and not to state a fact". This disagreement can be more or less extensive, and can concentrate on A or one of its intermediaries. In order to interrupt a translation, B can challenge A or I, either explicitly or implicitly. At the other end of the continuum, the translation may be accepted to the degree that it simply effaces itself, disappearing in the form of a constructed relationship and negotiated compromise. All that remains is self-evident agreement. There is clear empathy, a perfect piece of information that circulates without difficulty. Between these two extremes lie all those situations, so well described by game theory, in which A endeavours to anticipate what B wants and is thinking; wherein each puts themselves in the other's place and where translations sometimes succeed in stabilizing after a long series of iterations and speculations.. The successful translation creates the missing shared space, equivalence and commensurability: it aligns. Whereas when it fails, A and B return to an inability to communicate, and through a process of non-alignment reconfigure themselves in spaces without a common measure. What is central to the analysis is that the translation flows through the intermediaries *and* is held in place by them.[13]

When the translation is "perfect", what A says about A, I and B is the same as what B says about A, I and B or what I says about A, I and B. The equivalence is total, and the discourses are perfectly superimposible, without any ambiguity. The more one diverges from agreement, the more there are differences and incoherences. A no longer talks of I in the terms used by B; the definition of B given by A does not coincide with the one that B gives of itself, etc. In the former instance there is isotrophy, in the latter the space created is full of discontinuities; the range is from harmony to polyphony to cacophony.

A network begins to be constituted as soon as three actors A, B and C are aligned (by intermediaries interposed). There are two possible basic configurations for this alignment (Fig. 4.1). The first corresponds to a situation of complementarity (which results from the transitiveness of the relationships: A translates B, who translates C; therefore A translates C); the second, to one of substitutibility (A translates B who is also translated by C, who gives a similar definition of B). The degree of alignment depends on the degree of success of the translations (and in the case of substitutibility, on the extent of their similarity). Equally, this property which holds for three elements (A, B, C) enables one to identify sets of relationships obtained by composition of translations (since a chain of translations is itself a translation). The network is constructed according to the translations' own logic. Aggregation is not a procedure invented by the observer in order to simplify the complexity of reality, it is the very moment of social life. Whatever its length and degree of complexity, it is possible to gauge, at least qualitatively, the degree of alignment of any network of translations. We will speak of strong alignment when at any point the translations align the actors (whatever A and X one chooses, either there is a chain of translations such that A translates X, or there is a C such that A and X translate C in the same terms); it will be weak in the inverse case.[14]

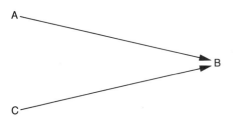

Figure 4.1

Any translation operation is accompanied by an attribution process, as a result of which the intermediaries in circulation are imputed to groups, who are thus transformed into actors. These attributions are a function of the play of interactions, which are — at least partially — codified into rules or conventions. These latter are at once the product of past interactions and a possible regulatory principle for present and future ones. Of course these conventions can be implicit or explicit, and they are always revisable and evolving.

What categories of conventions apply to the translation A→I→B?

The first set of rules deals with A's identity as an actor. Is A justified in claiming to be an actor? Is A really an actor? In other words, can intermediaries be attributed to A? Here there is a whole universe of conventions, from written law to custom. These conventions traverse the different poles of the TENs. Let us give a few examples: the definition of a legal entity, which allows attribution to a company of the products that it puts into circulation; legislation regarding industrial property, which can deny inventors the right to exploit their own inventions, these being attributed to the company paying the inventor's salary; or the unwritten law that prohibits the person funding a research programme from cosigning the articles that emerge from it.

The second set of rules concerns the attribution of a given series of intermediaries to a particular actor. A may be actor A, but this does not necessarily mean getting the attribution of I. This attribution is a function of A's skill. It also depends on a certain number of more or less complicated, explicit and transgressible conventions.

Let us assume A is actor A and has gained the attribution of I. As a result A can speak in the name of B. But this right can always be contested by B or by any other actor. Here we come across a third set of regulations, which codify the space of possible denunciations of A. This is particularly explicit in the political sphere, with its legitimated procedures for designating representatives, and also in industry, in the form of contractual agreements or collective accords which detail the hierarchy of responsibilities and working conditions.

All these conventions produce the same result, which is (more or less strictly or negotiably) to constrain the universe of possible actors by organizing attribution and limiting the number of stabilizable translations. I propose to call these regulatory processes forms of coordination. They codify translations, but only up to a point, and with varying degrees of durability, constraints and openness to change. In TENs — which include researchers, companies, users, technologists — several forms of coordination can be in operation simultaneously (the market, the organization, confidence, recognition . . .). Any one of them can be seen as a specific set of conventions defining the translation regimes (authors' rights, attribution mechanisms, ability to speak in X's name etc.), as well as defining the particular categories of intermediary that serve as the medium for the translation. Since in this text I am not concerned with the content of any coordination, but with its role in the establishment and dynamics of translation, I will distinguish two extreme situations. These correspond to the classification proposed by Thevenot (1985) in his analysis of the degree of generality of investments of form. At one extreme can be found coordination which involves everyone. Its aim is to apply to anyone without distinction (the object of the convention, naturally, being to define this generality: a citizen, an official diploma, a guaranteed loan, standardization of a technical object, etc.). At the other can be found local coordination, which is limited in scope. That is to say that while it draws on more general conventions, it aims to parcel up the complete universe of intermediaries, actors and their relations into specified sub-sets (a network, or the pole of a network) outside of which these conventions lose their validity.

This abstract definition corresponds to easily identifiable realities. Here are a number of instances of regulations with limited scope: the constitution of a cartel, the establishment of a collective accord for an industrial sector, the adoption of qualification requirements, the creation of a technical norm limited to a few producers and users, the setting-up of a consumer group, the organization of a professional association or a scientific society, the constitution of cooperative research centres. In these cases, we are dealing with local coordination, which frequently presupposes the existence of a more general regulation (antitrust laws, laws covering associations etc.).

Evidently, one should not reify this distinction. It is clear that a rule's degree of generality is always relative and is the result of a process of construction. The most general convention is in principle liable to become local again if the denunciations or challenges that it submits to succeed in inverting the balances of forces in its favour (Reynaud, 1989). Further, it is always possible to imagine the progressive extension of local agreements (a convention from one sector imposing itself on the whole economy, a private norm that becomes public, conditions for guaranteeing credit becoming generalized, etc.). We will use the term "weak coordination" to characterize a network which has not added on any local rules and procedures to conventions generally followed at any given moment. We will call the coordination strong in the inverse case. When the coordination is strong, the universe of translations is constrained and the networks become more predictable. When it is weak, it is less constrained and there are many possible developments and associations.

Let us call a network's degree of convergence the combined index resulting from its degrees of alignment and coordination. This concept rests on the simple idea that the more aligned and coordinated a network is, the more the actors composing it work together in a common enterprise, without their status as actors being under constant challenge. This does not mean that everyone does the same thing (remember that these networks can include researchers, technologists, entrepreneurs, salespeople and users). Rather it quite simply means that any one actor's activities fit in easily with those of the other actors, despite their heterogeneity.

To further illustrate the meaning of a convergent network, we could say that any actor belonging to the network, whatever their position (researcher, engineer, salesperson, user, etc.) can at any time identify and mobilize all the network's skills without having to get involved in costly adaptations, translations or decoding. The network as a whole is behind any one of the actors who makes it up. Faced with an angry client, for instance, the salesperson immediately knows which engineer to call on, how to formulate the problem so that the said engineer can get to work right away and if necessary establish a connection with a basic researcher who can receive the message, appropriately reformulated. Then back from the laboratory, by way of a whole series of intermediaries and successive translations, will come recommendations, replies, measures and decisions, which will allow the salesperson to keep the client linked to the network. What holds in the one direction also holds in the other. In a highly convergent TEN, basic researchers are well aware of the fact that the problems they are set coincide with a network of expectations and demands ready to take up their results once they emerge from the laboratory. The world surrounding• the researchers has long been prepared in such a way that the position of the laboratory and its research topics are closely tied to what the other actors do, want and expect. A totally convergent network would be a sort of Tower of Babel, in which each would speak in their own language and everyone else would understand them, and each would have skills that all the others would know how best to use. Such a network would be particularly efficient, since it would dispose at once of the force of the collective and the synthetic capacity of the individual. Any particular actor would be able to speak in the name of all, to mobilize at a point all the skills and alliances in the network.

Such a network, capable of concentrating itself in a point at the same time as deploying itself simultaneously in the several environments of science, technology, industry and consumption, is of course an exception, a limiting case.[15] By way of contrast, we should complete the range of possibilities by looking at another extreme situation: that of the very weakly convergent network, where it is difficult for a given actor to get recognition as an actor, or to mobilize the rest of the network; but where the network is sufficiently in place so that this mobilization is possible, albeit with great difficulty.

These examples demonstrate that the construction of convergent networks presupposes long periods of investment, intense effort and coordination.

Boundaries

A network's boundary can be related to its degree of convergence. We will say that element Y is outside of a network if locating the links between it and the actors (A, B, C . . .) within the network significantly decreases the network's degree of convergence: alignment and co-ordination are weakened by the new translations which are required.[16] A possible objection here is that it might be difficult to quantify these evaluations. How can one calculate a degree of convergence, giving it a numerical value without which it would be impossible to trace the border, to distinguish between an inside and an outside? This question relates back to the concrete methods that enable one to discern and describe translations, that is to say, to recover from them the different categories of intermediary the networks have ascribed them. Since any intermediary can be expressed in words or texts, the question then becomes how is it possible to analyse a variegated body of texts which define actors, their identity and their relationships? In fact the appropriate algorithm is extremely simple, even if it presupposes an enormous amount of computation. The translation of B by A has succeeded in direct relation to the number of texts or textualizations in which the definitions of A and of B, as well as the relations between them, coincide (all statements of the form ArB being identical). As co-word analysis (which prefigures the necessary software) shows, this computation is not beyond the bounds of possibility, and reasonable approximations can be developed without too much difficulty (Callon *et al.*, 1986).

The question of boundaries leads me to distinguish between two types of TEN: long and short networks. The former are those which include the whole set of poles and intermediaries enumerated above. In particular, they extend up from basic academic research. The corresponding industrial sectors are those which economists call "science-based". Short networks do not go so far back. Even if there are occasional encounters with fundamental research, these links are not stable or systematic. The network is basically organized around the technical and market poles. This distinction measures the length of the detour that has to be organized in order to create or develop a market. In certain cases it extends right into laboratories doing fundamental research. In other instances, it does not go beyond the world of technology. Whether they be short or long, TENs have one basic property in common — that of being able to encourage and organize interactions between the different activities that they coordinate (Gaffard, 1989).

Irreversibilization
The concept of translation leads on to that of irreversibility. The irreversibility of a translation depends on the impossibility it creates of returning to a situation in which it was only one possible option amongst others, and also on the predetermination of later translations. This definition does not prohibit one from talking about degrees of irreversibility. The translation A→I→B more or less definitively eliminates a greater or smaller proportion of competing translations. It more or less strongly predetermines future translations and, in particular, the actors' identities. Thus defined, the irreversibility of a translation is not a property that the observer can measure independently. It is a relational characteristic, which is only actualized when put to the test. The impossibility for other (past or future) translations to develop and impose themselves is a battle, a fight that is never definitively won, and whose outcome depends on the actors in play.

How can a translation succeed in resisting the repeated and obstinate assaults of competing translations, and in the end eliminate them without any possibility of their return? The answer to this question lies in two words: durability and robustness. These properties, which can only be measured in the fray, are above all properties of intermediaries, of translation operators. It is possible to picture all possible degrees of resistance; their degree of robustness will depend on the extent to which the identities of A and B as inscribed in the translation themselves become resistant. In effect, A and B are groups which hold together more or less well. They are hybrid collectives, constantly threatened by dissension and internal crisis. They will be better protected from these questions and from dismembering (and this holds for a plant, a union, or a team of skilled workers or researchers) to the extent that their constituent elements are strongly associated. We must, however, be prudent in our description of the mechanisms by which translations impose themselves and eliminate their competition. No strategy is a priori assured of victory. Generally speaking, it could be said that irreversibility increases in proportion to the degree that systemic effects are created in which each element translated, each intermediary and each translator, is inscribed in a bundle of interrelationships. In this case, modifying any one element, that is to say defining it differently, presupposes that one engages in a process of generalized retranslation. We will venture the following proposition: the more heterogeneous and numerous the interrelationships, the stronger is the coordination and the higher the probability of successful resistance to alternative translations.

Whatever the durability and robustness of a translation, it says nothing about the extent of the predetermination of future translations. To what extent does a scientific text which "translates" a monoclonal antibody, by setting down its attributes, and which resists the most ferocious attacks, render a given research strategy and certain industrial developments necessary? To what extent does a micro-computer and its software, which assign its users precise roles and which define at the same time as they hierarchize the problems that can be treated, render certain behaviour and certain operations predictable? One might say that a translation is irreversible to the extent that it renders alternative translations improbable. The concept of apprenticeship (or learning) is central for analysing such a situation. Apprenticeship designates the set of mechanisms by which, through progressive mutual adaptation, the different elements involved in a translation (A, I and B) become exclusively dependent on each other. B cannot work except with machine tool I. Such and such a technical object cannot be developed except by specialists who have received particular training. A's trade binds him or her to putting I into circulation. And so on. Thus decisions are dependent on the history of past translations.

The creation of systemic effects and the apprenticeship process refer back to a more fundamental mechanism: that of the normalization of behaviour which accompanies (and measures) the irreversibilization of the translation A→I→B. As David (1987) notes, this process applies to all categories of groups, which can involve various degrees of association between humans, non-humans, texts and money. The functions of normalization are to render a series of links predictable, to limit fluctuations, align actors and intermediaries, and to cut down on the number of translations and the amount of information that they put into circulation. Normalization operates through the standardization of the different classes of interface: actors/intermediaries, intermediaries/ intermediaries, intermediaries/actors. This normalization is more or less constraining — ranging from reference standards to fully compatible interfaces by way of the definition of maximum and minimum thresholds. If the relationship A-I-B is normalized, it can contribute powerfully to the production of systemic effects. The elements that constitute it can only rearrange themselves with well-defined elements which adopt the same or compatible standards. The stricter the compatibility rules, the greater the permanence with which alternative translations are disqualified. A further effect of the immense normali-zation effort is predictability. (an almost null probability that A-B will be replaced by A-C). A network whose interfaces have all been standardized transforms all the actors composing it into docile agents, and all the intermediaries which circulate into stimuli which auto-matically evoke and sometimes determine certain types of response. The rules of coordination then become constraining norms, which create deviance at the same time as they control it: the past engages the future. In a word, irreversibilization, taken as the predetermination of translations and as the impossibility of a return to competing translations, is synonymous with normalization.

Any talk of normalization or standardization evokes the possibility of some quantification, however rudimentary. Imposing norms for inter-faces involves identifying at least one pertinent variable which takes one of two values (that is to say, good/bad; this one passes/this one does not). It can perhaps extend to involve fine tuning between multiple continuous variables, by way of fixing upper and lower limits for thresholds.[17] The more precise and quantified these standards are (that is to say, the more the actors and intermediaries are objects of a precise, known, and agreed characterization) the more the translation gains in irreversibility when it succeeds. A network which irreversibilizes itself is a network that has become heavy with norms of all sorts, and which as a result slips into a codified metrology and information system. It is not hard to mathematize the description of such a network, since the functioning of each element is quantitatively linked, by its specifications, to the various other elements in the network. For example, it is known how to associate the performance level of a technical object (speed, memory and power of a microprocessor), the category of user concerned and the price they are willing to pay. The translation can be given in the form of a table of correlations between heterogeneous numerical variables (if you can get up to 10 MHz, then the desktop publishing market opens up and you can be looking at prices over £5000, etc). These correlations can affect all or part of a TEN, and the different elements that make it up. A complementary approach is the techno-metric reduction (Saviotti and Metcalfe, 1984) of a technical object to a configuration of parameters describing the principal performances and their different uses, which can take a set of interlinked values. Take as an example a heat exchanger whose thermal interchange index lies between x and y, and which can be used for the two unconnected uses of the drying of grain or the recovery of heat in a crematorium. This general approach can be used to explore TENs. The same holds for the concept of the productivity of incorporated skills and their measurement. For example, it may be that a network's evolution leads to its changing the performance of a technical object (the memory of a micro, the speed of a telephone exchange) from value X1 to X3, and the quantity of incorporated skills from Q1 to Q2. One can give a good description of the network's dynamics by mapping a few parameters around which translations are organized and crystallize themselves (Rabeharisoa, 1990).

With the irreversibilization of the translation and the normalization it leads to, we enter into a world familiar to economists (Akrich, 1989c). In effect, it becomes possible to say that it would be expensive to challenge one or several translations which not only resist competing translations but also restrict the number of possible future translations. This means that in order to establish other links and set up new translations you would first have to undo those which already exist, and change the equivalences in operation, which would in turn mean mobilizing and enrolling new alliances. Economics does not begin with the allocation of rare resources, but with their localization. Thus non-linearity and path dependency can be seen to be integral to the dynamics of the economy.

FOLLOWING THE DYNAMICS OF NETWORKS

We have seen that a TEN can become shorter or longer (to include actors from the scientific pole); it irreversibilizes or reversibilizes; its convergence increases or decreases. A network's internal dynamics can be related to these three dimensions, which constitute coordinates enabling us to identify the trajectories followed by a network, and to describe them. A simple four-quadrant diagram will enable us to depict these trajectories, the description of whose dynamic is dependent on the degree of information one provides about the network's length (Fig. 4.2).

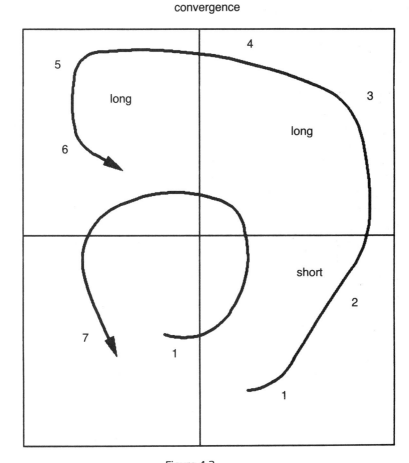

Figure 4.2

The greater the extent to which the trajectory traverses the different quadrants in the diagram, the more frequent are the changes in length of the network, and the more profound are its transformations. Indeed, when the translations vary, the content of the intermediaries that circulate, the identity of the actors involved and the morphology of their relationships all vary. In this case, not only are technical objects, scientific texts, incorporated skills and the flow of money subject to vast change and unpredictability, but also the identity of the actors. When the network arrives at or remains within the strong irreversibility/strong convergence quadrant, then the translations intermingle and inscribe themselves in stabilized intermediaries. All the actors, and the intermediaries through which they define thesmelves and others, maintain the same identity. In particular, strong convergence implies formalized coordination — that is to say, the existence of numerous conventions and local procedures which create that strange situation in which human beings and technical objects evolve predictably, as if acted on by rules to which they conform. The concepts of "routine" or "heuristic" apply perfectly to this particular configuration (Nelson and Winter, 1977; Van der Belt and Rip, 1987). Our analytical framework enables us to chart the evolutionary process which sometimes — but this is not at all obligatory or fixed for all time — sees a network pass from a state of flux and divergence to one of strong irreversibilization and standardization. Thus we find mechanisms governing the use of a particular pole of the network inserted into a more general framework. For example, if a network has succeeded in aligning buyers (the market pole) defined by their demand for a standardized product, then it is possible to speak of a combined demand curve, analysable as representing any one of the consumers, between which the network has constructed perfect equivalence. If the network has created complete convergence between researchers (scientific pole), then the Kuhnian paradigm which results can be described as a unit, a single research programme. In the same way, the product cycle of a standard model represents the particular trajectory of a network in which the alignment between companies, technicians and buyers is so perfect that the evolution of the whole occurs in phase, without any hitches. But analysis in terms of TENs is infinitely more tolerant than more deterministic models. It enables one to chart the most heterogeneous convergences, such as those linking scientific actors to technological ones, in order to compose coordinated groups in which science and technology interact closely. Within a single network, it enables us to distinguish between sub-sets having different degrees of convergence. Finally, it does not impose any a priori form of evolution. Convergence can increase or decrease, and the same holds for irreversibility.

The above shows that, unlike trajectories, networks can rarely be cut up into simple and easily quantifiable descriptive frameworks. Putting things into numbers, the extreme case of putting things into words, is only one form of description among others, the use of which clearly depends on the state of the network. It would make no sense to try to quantify at any price, or to seek to reduce behaviour to variables or functions, just as it would be silly to reject all quantification out of hand. The choice of method does not obey any epistemological imperative — it falls outside the province of any doctrine, since it is entirely dictated by the state of the network. If the network "standardizes" itself, then one has to count and do some mathematics. If it is divergent and reversible and remains in this quadrant, then any excessive simplification (and as a consequence any quantification) risks betraying the state of the network and its dynamics. Without doubt in this case it would be better just to tell the story! Indeed in this case each actor is relatively unpredictable, because any translation that it attempts is constantly undone. In this case, the only method that enables one to faithfully and intelligibly account for what happens is a literary description which multiplies the points of view, forming a polyphonic narrative distributed over as many voices as there are actors and recovering all the attendant details.

When a network is strongly convergent and strongly irreversibilized, it can be approximated as a black box, whose behaviour can be known and predicted independently of its content. It can then link itself to one or several "external" actor-networks, with which it exchanges inter-mediaries, which traverse its border in both directions. We will say that it punctualizes itself in other networks, whose dynamic it participates in (Callon, 1987). The relationships between a network-point and the networks surrounding it can be analysed in terms of translation, the intermediaries which circulate being its operators. Punctualization can also apply to a complete industrial sector (the microprocessor industry, a black box producing certain categories of product with well-defined characteristics and consuming particular categories of input). It can also apply to a scientific discipline, a technological sector, or a market.

This process of punctualization, which folds up an entire network so as to transform it into a point in another network, which at the same time becomes more general and more encompassing, is the basis of what is called clustering, or the progressive passage from the micro to the macro. Thus we define clustering in terms of network closure, by the constitution of black boxes which juxtapose themselves with other black boxes (that is to say with other punctualized networks) by linking to them by way of translation operations no different from those examined already. Moreover, the network-point can in turn act as an actor or an intermediary. Thus we also have disaggregation proceeding from the opening of black boxes, the redeployment of punctualized networks.

In these cases, the degrees of convergence and irreversibility decrease catastrophically — markets collapse, industrial sectors are dislocated, scientific specialities disintegrate, etc.). The overall dynamic should follow these foldings and unfoldings. Where one is placed is without doubt central for describing these different configurations.

ACTORS, INTERMEDIARIES AND NETWORKS

As defined above, TENs are not like networks as normally defined. They bear only a distant family resemblance to the technical networks normally studied by economists (telecommunications networks, train networks, sewers, etc.) which can essentially be reduced to long associations of non-humans that here and there link a few humans to each other. Nor are they reducible to the networks of actors described by sociologists, which privilege interactions between humans in the absence of any material support. Techno-economic networks are composite. They mix humans and non-humans, inscriptions of all sorts, and money in all its forms. Their dynamic can only be understood by way of the translation operation which inscribes the mutual definition of the actors in the intermediaries which are put into circulation. Knowledge of these networks involves "reading" these inscriptions. Further, the translation operation is itself regulated by conventions that are more or less local, and are always revisable.

One of the advantages of reasoning in terms of TENs is that it shows that an actor's theory can in no way be universal. The behaviour of the actors, and more generally their definition, changes with the state of the network, which is itself the result of previous actions. It should be possible to characterize the actors and their action profiles for each possible configuration of a network, along the three dimensions of length, convergence and irreversibility. The less convergent a network, the less it is irreversibilized and the more the actors composing it can be understood in terms of concepts like strategies, variable and negotiated aims, revisable projects and changing coalitions. At the other extreme, in completely convergent, irreversibilized networks, the actors become agents who have precise objectives. The states of the network are known for each point at each instant. Information as delivered by the translation inscribed in the intermediaries is at once perfect (the network is known and predictable) and limited (it does not go beyond the network under consideration). Moral risks and adverse selection (to use the language of the economist) or controversies and *desinteressement* (in the repertoire of translation sociologists) are highly improbable. The paradox is that in these situations of perfect information, the actors are incapable of choosing, since they are "acted" by the network holding them in place; they are only in a position to act deliberately when there is imperfect and asymmetrical information.[18] There are many intermediate situations between these two extremes — such as, for example, that of limited or procedural rationality, or that of mutual anticipations in game theory (Thevenot, 1989). This line of analysis is worthy of development. If it proves to be well founded, it opens up an entirely new space for the social sciences. It suggests that there is no theory or model of the actor. The actor's ontology has a variable geometry, and is indissociable from the networks defining it, and which it, with others, contributes to defining. The historical dimension becomes a necessary part of the analysis.

Some will say that we have proposed a method for describing TENs, but not a theoretical framework enabling one to explain their functioning. This common opposition between description and explanation is in large part thrown into doubt by the procedure we propose. The more a network increases its degree of convergence and irreversibility, the more the descriptions that the intermediaries in circulation deliver become explanations, or even predictions. Talk of explanations presupposes that it is possible to account for the state of a network and its evolution from the basis of a small number of variables or concepts. This involves making a very definite hypothesis about the form of the network and the convergence of its translations. In a strongly convergent and irreversibilized network, the actors are perfectly identifiable, and their behaviour is known and predictable. The whole works and evolves according to regularities which enable one to explain the trajectories followed, the division of resources and the equilibria attained, on the basis of a few simple laws and some well-chosen information. In a divergent and reversible network the description has to include every detail; each actor endeavours to translate the others, and these translations fluctuate without ever succeeding in stabilizing. Anyone looking for explanation would not understand anything in terms of the mechanisms by which irreversibility is created, and would be incapable of saying anything sensible about the state of the network and its transformations. Those who pit qualitative analyses, monographs, and strategical or prospective analyses against the search for laws and regularities quite simply overlook the fact that networks are not *in* the actors, but are produced by them, and that they only stabilize at certain places and at certain times.

Notes

1   See among others Callon and Latour (1981), Callon (1989), Dosi (1984), Freeman (1982), Gaffard (1989), Hughes (1983), Kline and Rosenberg (1986), Latour (1989), Von Hippel (1988).

2   Here I refer back to another article on the description of flexible techno-economic networks, which constitute one of the dominant forms of contemporary industrial activity. For a characterization of the morphology of TENs, see Callon *et al.* (1990).

3   These activities of intermediation are fairly similar to the compromises between natures described by Boltanski and Thévenot (1987).

4   As we will see below, the distinction between intermediaries and actors has to be manipulated with care, since they are often one and the same.

5   The solution that I propose for establishing a bridge between sociology and economics is different from that which comes out of the notion of "embeddedness" revived by M. Granovetter (1985). The networks he describes, which are pure associations between human beings, are very different from TENs.

6   For the idea of immutability, which is essential for understanding action at a distance, see Latour (1989).

7   There is a discipline — scientometrics — which is entirely devoted to the decoding of the inscriptions provided by articles.

8   For an analysis showing how patents can be decoded, see Bowker (1989).

9   There are many unexpected turns and slips in the design of a technical work as in the discourse of someone under psychoanalysis — and users play on this continually. Hence the importance of the idea, dear to economists, of learning by using.

10   Equally, textualization occurs when the object gives rise to controversies, alternative and contradictory network formations. The descriptions (in the form of accusations) proposed by the different protagonists are a chaotic mixture of the technical and the social.

11   Suppose that our person on the analyst's couch in the earlier footnote is not considered as the subject to whom his or her discourse can be attributed (this case is not at all fantastic: psychoanalytic treatment affirms from the beginning that it is "that", and not the subject who speaks; just as exorcism seeks to uncover Satan). In the same movement, the actor shifts. The person under treatment is only the medium chosen by the unconscious to express itself, becoming a set of symptoms to decode. The person at the confessional no longer has free will, but is possessed by the devil. As can be seen from these examples, the observer does not have to oppose his or her interpretation to that of the actor in order to denounce appeals to the unconscious in order, for example, to impute to the person being analysed the responsibility for his or her intermediaries.

12   The intermediary (pure or hybrid) is what we have called elsewhere the translation operator, or again the apparatus for arousing interest (*dispositif d'intéressement*).

13   It could be shown that machines, human bodies and texts, considered as intermediaries, are at once the basis of all possible renderings, misunderstanding, and also of all (re)conciliations (the telephone creates a common space that integrates as much as Durkheim's religion or Bourdieu's habitus; nuclear plants generate conflicts as intense as those over the rights of man).

14   We should deal with the question of how the notion of alignment applies to the different poles of a TEN. Take the case of the market pole. The users will be aligned either if they all ask for the same standardized product (substitutibility) or if the choice made by each of them is mechanically linked to the choice made by the others (complementarity). In the first configuration, we recover the neoclassical orthodox model, and in the second a situation close to that described by the sociology of consumption or by the economy of network externalities. We should add that in order to analyse market structures, as defined by economists (confrontation between a supply and a demand), we have to add to the description of our market pole a description of the technical and market poles. There are then a great many more possible configurations, and it is easy to show that one can recover the main structures already known, and uncover new ones.

15   The archetype of the TEN in which relations are formed without any discontinuity from science to the market-place is furnished by the nice study by Bêta's team of material science. At one extremity, there are users who express their demands in terms of functionalities they want from the material (gluable, able to be soldered, heat resistant, light, capable of absorbing such and such a degree of mechanical pressure). At the other, there is basic research on the physical microstructures involved in ensuring that the materials have the given combination of properties desired. Between the two there are materials made to measure, and flexible production systems enabling optimization of the production of ranges of different products, research efforts on generic technologies like soldering, gluing, collaborations, alliances, research contentions etc. In brief, there is an almost direct relationship from the most basic research to the user — one, however, which passes through a series of intermediary stages which have been carefully articulated with each other (Cohendet et al., 1987).

16   As can be seen, this definition is different from that used in classic clustering algorithms, which trace the limits of clusters as a function of a threshold imposed on the intensity of the relationships between elements. What is determinant is the degree of convergence, and not the intensity of a given relationship.

17   It is not difficult to give examples of such standardizations, which link all the classes of possible groups:

(a)   In the case of groups made up mostly of humans, one can speak — following Riveline (1983) and Oury (1983) — of management parameters for indicating the existence of norms generally calculated by regulating the behaviour of certain agents and describing their relationships: in order to stay in the network, a salesman has to contact more than 20 potential clients per month (definition of a minimal threshold), the production engineer should not have more than x rejects (maximum threshold), freelance journalist Z's payment (measurement of the relative attachment) is proportional to the number of lines written . . . .

(b) Norms between non-humans (called technical norms): the sub-system disconnects itself if the current exceeds a certain value (fuse); you cannot plug something in unless the pins and the plug match, or unless the voltage is within 5 per cent of the optimum.

(c) Human/non-human norms: if the signal light blinks, the operator has to press a lever (which is of the form: if the pressure exceeds a certain threshold then such and such an action will be begun).

(d) Norms organizing the relationships between scientific texts: inscription of the journal's name on each page in the article, standardization of references and diagrams.

18 Dupuy (1989) develops a similar argument. This could be formulated differently: the existence of the neo-classical market presupposes in reality the existence of strong alignments (notably users/clients).

Acknowledgements

This text is in large part the fruit of discussions with all my colleagues at the CSI, and particularly of the dialogue I have been able to maintain over several years with Bruno Latour. I would like to thank the following for their valuable comments: L. Boltanski, G. Bowker, D. Fixari, A. Hatchuel, J. Law, C. Riveline, A. Rip, L. Star and L. Thevenot.

References

Aglietta, M. and Orlean, A. (1982), *La violence et la monnaie*, PUF.
Akrich, M., Callon, C. and Latour, B. (1987), A quoi tient le succes des innovations, *Gérer et Comprendre*, **11 and 12**.
Akrich, M. (1988), Comment décrire les objets techniques *Technique et Culture*, **9**.
Akrich, M. (1989a) De la position relative des localités; systèmes électriques et réseaux socio-politiques, in *Innovation et ressources locales*, Cahiers du CEE, **32**.
Akrich, M. (1989b), La construction d'un système socio-technique: esquisse pour une anthropologie des techniques, *Anthropologie et Société*, **12**.
Akrich, M. (1989c), "Essay in technosociology, a gazogène in Costa Rica", in P. Lemmonier (ed.).
Akrich, M. and Boullier, D. (1989), *Représentation de l'utilisateur final et genèse des modes d'emploi*, LARES-CCETT.
Arthur, B. (1989), Competing technologies increasing returns and lock-in by historical events, *The Economics Journal*, March.
Bijker, W. and Law, J. (eds), (1990) *Constructing Networks and Systems: Case Studies and Concepts in the New Technologies Studies*, MIT Press, Cambridge, MA (in press).

Bijker, W.E., Hughes, T.P., and Pinch, T. (1987), *The Social Construction of Technological Systems: New Directions in the Sociology and History of Technology*, MIT Press, Cambridge, MA.

Boltanski, L. and Thevenot, L. (1987), *Les économies de la grandeur*, CEE-PUF.

Bourdieu, P. (1980) *Le sens pratique*, Editions de Minuit, Paris.

Bowker, G. (1989), *What's in a patent*, CSI.

Callon, M. (1976), "L'opération de traduction comme relation symbolique", in *Incidences des rapports sociaux sur le développement des sciences et des techniques* (Ed. P. Roqueplo), CORDES.

Callon, M. (1980), "Struggles and Negotiation to Define what is problematic and what is not. The Socio-logics of Translation", in K.D. Knorr, R. Krohn and R.D. Whitley.

Callon, M. (1986) Eléments pour une sociologie de la traduction, la domestication des coquilles St Jacques et des marins-pêcheurs dans la baie de St Brieuc, *L'Année Sociologique*.

Callon, M. (1987), "Society in the making" in Bijker *et al.*

Callon, M. (1989), *La science et ses réseaux*, La Découverte.

Callon, M., Laredo, P., Rabeharisoa, V. Gonard, T. and Leray, T. (1989) "Des outils pour la gestion des programmes technologiques: le cas de l'AFME", in D. Foray.

Callon, M. and Latour, B. (1981), "Unscrewing the Big Leviathan: How Actors macro-structure Reality and How Sociologists Help them to do so", in *Advances in Social Theory and Methodology: Toward an Integration of Micro and Macro-sociologies*, (Eds K.D. Knorr-Cetina and A.V. Cicourel), Routledge and Kegan Paul, London.

Callon, M. and Law, J. (1982), On Interests and their transformation, *Social Studies of Science*, **12**.

Callon, M., Law, J. and Rip, A. (1986) *Mapping the Dynamics of Science and Technology*, Macmillan, London.

Cambrosio, A. and Limoges, C. (1990) "The controversies over the environmental release of genetically engineered organisms: shifting cognitive and institutional boundaries".

Cohendet P., Ledoux, M. and Zuscovitch, E. (1987) *Les matériaux nouveaux: dynamique éconmique et stratégie européenne*, Economica.

Crozier, M. and Friedberg, E. (1977) *L'acteur et le système*, Le Seuil.

David, P. (1987) "New standards for the economics of standardization", in *Economic theory and technology policy*, (Eds Dasgupta and Stoneman), Cambridge University Press, Cambridge.

Deleuze, G. (1989) *Le pli*, Editions de Minuit.

Dosi, G. (1984) "Technology and conditions of macroeconomic development", in *Design, innovation and long cycle in economic development*, (Ed. C. Freeman), Frances Pinter, London.

Dupuy, J.P. (1989) Convention and common knowledge, *Revue Economique*, **2**.

Foray, D. (1989) Les modèles de compétition technologique: une revue de la littérature, *Revue d'Economie Industrielle*, **48**.

Freeman, C., (1982) *The Economics of Industrial Innovation*, Frances Pinter, London.

Gaffard, J.L. (1989) Marché et organisation dans les stratégies technologiques des firmes industrielles, *Revue d'Economie Industrielle*, **48**.

Gille, B. (1978) *Histoire des techniques*, Gallimard, Paris.

Granovetter, M. (1985) Economic action and social structure: the problem of embeddedness, *AJS*, **91**(3).

Hughes, T. (1983) *Networks of Power: Electrification in Western Society, 1880–1930*, The John Hopkins University Press, Baltimore, MD.

Katz, M. and Shapiro, C. (1985) Network externalities, competition and compatibility, *American Economic Review*, **75**.

Katz, M. and Shapiro, C. (1986), "Technology adoption in the presence of network externalities", *Journal of Political Economy*, **94**, 4.

Kline, S. and Rosenberg, N. (1986) "An overview of innovation", in *The positive sum strategy*, (Eds R. Landau and N. Rosenberg), Academy of Engineering Press.

Knorr, K.D., Krohn, R. and Whitley, R. (eds), (1980), *The Social Process of Scientific Investigation, Sociology of the Sciences Yearbook*, **4**, Reidel, Dordrecht.

Latour, B. (1984), *Microbes: guerre et paix*, A.M. M taili.

Latour, B. (1988a), Mixing Humans and Non-Humans Together: the Sociology of a Door-closer, *Social Problems*, **35**.

Latour, B. (1988b), *La vie de laboratoire*, La Découverte.

Latour, B. (1989), *La science en action*, La Découverte.

Latour, B. (1991), *Aramis ou l'amour e la technique* (in press).

Latour, B. and Coutouzis, M. (1986) Le village solaire de Frangocastello: vers une ethnographie des techniques comtemporaines, *L'Année Sociologique*.

Law, J. and Callon, M. (1988) Engineering and Sociology in a Military Aircraft Project. A Network Analysis of Technical Change, *Social Problems*, **35**.

Law, J. (1986) *Power, Action and Belief: A New Sociology of Knowledge?*, Routledge, London.

Law, J. (1987) "Technology and Heterogeneous Engineering: the Case of Portuguese Expansion" in: Bijker, W., Hughes, T.P. and Pinch, T. (eds).

MacKenzie, D. and Wajcman, J. (1985), *The social shaping of technology*, Open University Press, Buckingham.

Mustar, P. (1989), *La création d'entreprises par des chercheurs: deux études de cas*, CSI.

Nelson, R. and Winter, S. (1977), "In search of a useful theory of innovation", *Research Policy*, **6**.

Oury, J.M. (1983), *Economie politique de la vigilance*, Calmann Lévy.

Parsons, T. (1977), *The evolution of societies*, Prentice-Hall, Englewood Cliffs, NJ.

Rabeharisoa, V. (1990), *La construction de réseaux technico-économiques dans le domaine de la régulation thermique*, CSI-AFME.

Reynaud, J.D. (1989), *Les règles du jeu*, A. Colin.

Riveline, C. (1983), Nouvelles approches des processus de décision: les apports de la recherche en gestion, *Futuribles*, **72**.

Saviotti, P. and Metcalfe, J.S. (1984), A theoretical approach to the construction of technological indicators, *Research Policy*, **13**.

Star, S.L. (1988), Introduction: The Sociology of Science and Technology, *Social Problems*, **35**.

Thévenot, L. (1985), "Les investissments de forme" in *Conventions économiques*, CEE-PUF.

Thévenot, L. (1989) " Equilibre et rationalité dans un univers complexe", *Revue Economique*, **40**.

Touraine, A. (1974), *La production de la société*, Le Seuil.

Van der Belt, H. and Rip, A. (1987), "The Nelson-Winter-Dosi model and synthetic dye chemistry", in Bijker *et al.* (eds).

Von Hippel, E. (1988), *The sources of innovation*, Oxford University Press, Oxford.

5

PARADIGMATIC CHANGE, NORMATIVE UNCERTAINTY AND THE CONTROL OF KNOWLEDGE

Ray Loveridge

INTRODUCTION

Most overarching theories of innovation express the learning process in the S-shaped curve form adopted by Rogers in 1962. Like Rogers himself (1983), subsequent analysts have usually broadened the base of the assumptions underlying the shape and linearity of this model. Nevertheless, macro-level studies can often be categorized as being concerned either with the process by which the invention/design of products moves from the idiosyncratic to more standardized forms, or from the locally isolated adoption of a good to its diffusion across a satiated market place. In both Europe and the USA, many of the pioneering studies of the in-firm adoption of innovative capital equipment (described over the last quarter century as New Technology) have been carried out by industrial sociologists working in the Labour Process tradition. Invention/design studies have more often been seen as the domain of institutional economists focused on sectoral boundaries or technology policy. To some extent the two approaches have been bridged within the contingent analysis of the so-called New Institutional economics or transaction-cost perspective (Williamson, 1985), as well as by more radical historical analysis (Piore and Sabel, 1984).

These different sets of scholars share frequently implicit assumptions about the nature of the strategic rationality exhibited in the formation and pursuit of corporate goals and strategies — whether seen to be the outcome of conscious or unconscious intent. Discontinuities in market and technological trajectories are often described as being accompanied by crises brought about by the ensuing disruption. This disruption may be seen as located within the structures of capital formation (Child and Loveridge, 1981) and/or within the cognitive structures or mental paradigms used to design and interpret their control by significant actors (Dosi, 1982; Perez, 1983).

In this chapter this rationality assumption is questioned in the light of differences in the relational bases or social frames which shape the actual operationalization of the "formulae" provided by shared techno-economic blueprints. It is suggested that crises are more likely to arise in situations inducing cognitive dissonance within strategic perceptual frames of reference. This is a more culturally specific interpretation than would be allowed by prevailing prescriptive theories of innovation based on Anglo-American experience.

THE CYCLICAL NATURE OF INNOVATION

The most common metaphor used in the explanation and prediction of innovation is the life cycle model. This is normally expressed in the form of the ubiquitous S-curve, usually denoting early difficulties in learning and diffusion of an innovation, followed by codification of design and operational techniques and the more rapid and/or widespread diffusion of a standardized "best practice". When improvements in performance in measured outputs is indicated on the Y axis, a similarly shaped curve has been used to denote improvements in a technological configuration attributable to a logically incremental learning process within a given firm or sector (Nelson and Winter, 1982; Georghiou et al., 1986).

The same metaphor is also particularly well developed in descriptions of the phased emergence of industrial boundaries. Abernathy et al.'s (1983) study of the automobile industry has provided a prototype for several derivative taxonomies using a stepped function to illustrate emergent market requirements and design templates. These move from the emergence of revolutionary new product designs within unstructured markets, to their maturity in design and finally to the niche creation symptomatic of very mature products and production methods. Shearman and Burrell (1988) describe a not entirely unrelated, four-stage process of the institutionalization of interfirm relations in a manner that contributes to the creation of an industrial or sectoral identity. Population ecology theorists also see the demands of interfirm networks as bringing about an "isomorphism" in the development of firms (Freeman and Barley, 1990) while a much stronger tradition in organizational sociology and economics attributes senescence in the "metabolism" of the individual firm to a growth in internal complexity and rigidity in procedure (Chandler, 1977; Kimberley, 1980). This trajectory has been seen to be punctuated by transitory, firm-specific crises which lead to shifts between different modes of administration.

The phases of adaptation can be seen to represent movements between the central integration of control, followed by its devolution to operating units and back again in sequence, as organizational complexity increases. Complexity in strategic decision making is often expressed as a function of increasing size of operations and diversification of goals. This staged sequence of modalities can be related to attempts at coping by successive managerial regimes (Child and Francis, 1979).

The cusp or discontinuity that sets the firm and industry on to a new innovation path is often illustrated by the interpolation of an overlapping curve, illustrating an emergent market or production process even as the old one reaches its zenith. The two overlapping processes are sometimes presented as interrelating only through a traumatic realization by the corporate entrepreneur of new competition. Abernathy *et al.* (1983) suggest that such crises accompany the simultaneous realization of obsolescence in market linkages and in design and production competences (and more generally, in the relational and informational exchanges with other actors making up the cognitive arena of the strategist (Child and Smith, 1987)). Minor crises of adaptation are seen to accompany problematic adjustments in one or the other of these exchanges during previous phases of development of the firm in its sector.

Revolutionary changes in design complemented by a transformation in market boundaries bring about the most profound state of "transcilience" through which institutions must pass in order to survive. This "architectural" phase is one in which the operational uncertainties created by the loss of markets or other network relationships is conjoined with an intellectual incomprehension of the new operational disciplines — for example, as when hot lead typesetting was replaced by computerized direct entry in newspaper printing and publication. At the same time, changes in distribution and a shift to financing through advertising transformed the former market basis of newspapers (Lee, 1982).

The phases of sectoral change are illustrated by Abernathy *et al.* (1983) in a two-dimensional diagram in which the process is described in terms of the effectiveness of the competences possessed by firms within the sector, and the strength of market linkages enjoyed with customers. This latter dimension might be expanded in the manner suggested within Fig. 5.1, to subsume all boundary-spanning relationships — both suppliers and customers — that provide the firm with its *raison d'etre*. Though these relationships may be assumed to be competitive at some level, there seems no reason to believe shifts in consumer demand to be the only source of debilitating change.

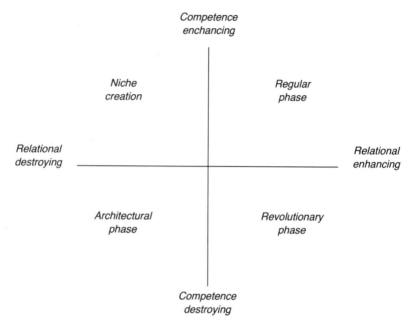

Figure 5.1: Phases in sectoral or corporate learning (derived from Abernathy *et al.* (1983), p. 10. See also Tushman and Anderson (1986)).

CRISIS TO TRANSFORMATION?

The crisis metaphor has found an easy validity in the context of the widespread collapse and renewal of Western manufacturing and service sectors over the last twenty years. In this setting it is unsurprising that there should have been a revival of interest in the Kondratiev/ Schumpeter (1939) explanation of long waves or cycles of economic activity in developing capitalist economies. Freeman's (1979, 1982, 1987) work has been most prominent in Britain in bringing together the interaction of market and inventive activities over (yet another) four-phase cycle of diffusion and technical development. Perhaps one of the more important elements in this enterprise is Freeman's attempt to explain the diffusion of innovation in a manner that allows his sectoral configurations to be related to the empirical categorization of inter-corporate relationships devised by Pavitt (1986).

Other theorists have attempted to interpret the crisis metaphor in terms much closer to its Marxian roots. Here Aglietta (1975) and the

French "regulation" school have been extremely influential in their categorization of historical movement between crises in "regimes of accumulation". Edwards (1979), and subsequently Gordon *et al.* (1982) have associated global economic crises with the degeneration of control over the labour process in key sectors of developing economies.

In both cases it may be noted that such events are precipitated by changes in the relational expectations or aspirations of key groups of actors, rather than simply by shifts in "objective" characteristics of economic structures. However, many global theorists in the same teleological tradition have placed considerable emphasis on the latter in offering a new prototype for the next epoch.

Perez (1983) hypothesizes that the new "technological style" or "techno-economic paradigm" is arrived at through a learning process that stretches over the several stages of development put forward by Freeman. This diffuses by moving along a path stretching across sectors arranged according to their functions in providing motive, carrier or induced markets for the new technology. The new cognitive paradigm that emerges represents a response to what is seen as the final state of "stable dynamic" within the relative cost structures across these sectors. It is arrived at through a process of trial and error, but above all it relates to the emergence of a virtually unlimited supply of a "key factor" which, historically, has taken the form of a new energy source or a new form of material transformation. Currently it is seen as applying to cheap information technology (IT). The pattern is seen to be supranational, because similar opportunities for entrepreneurs to use the new technology will present themselves to all and, apparently, will result in a similar global configuration in use patterns.

The description of the firm that emerges as the prototype for the new epoch does not greatly vary across different schools of analysis. In many ways it is no more than a stylization of fragmented observations of actual behaviour among multinational corporations over the last decade. Its characteristics include the dispersion and devolvement of operational control, functional and numerical flexibility in the use of labour, and the greater use of external contracts for both the provision of routinized services, and as the basis for temporary collaboration in high risk areas of the firm's activity, such as R and D.

One such model is that of "flexible specialization", put forward by Piore and Sabel (1984), as the post-Fordist solution to the failure of the highly centralized, functionally divided, Fordist models of servicing mass consumer markets. Flexible specialists are seen as achieving economies of scope through their ability to service several market niches, or more usually several supply chains to assemblers of consumer goods, within a narrow type of output. An equally important source of market leverage derives from the external economies achieved through location in industrial districts in which relations of competitive collaboration have created a collective infrastructure that complements individual activities in a manner that has given rise to a competitive advantage in global markets (Marshall, 1920). A similar model of mutual coordination of activities and "internal" dependencies is seen to exist within the structure of Keiretzu corporate groupings in Japan (Freeman, 1979; Best, 1990).

CRISIS OR CONTINUITY?

Critics of the transformation-crisis approach to innovation point to the continued role of long-lasting firms in the sponsorship and orchestration of research and design and marketing of new products (Pavitt, 1986). These are often multinational corporations whose influence extends to that of providing markets for flexible specialists and for "industrial districts" (Amin and Robins, 1990). This second body of thought suggests that radical innovation can usually be traced to an accumulation of incremental changes, and is often an extension of known processes and design to new groups of consumers. The total redundancy of a firm's competences is seen to be a rare event; more often, "learning by using" provides the basis for both new product launch and for the continued development of "robust designs" (Rothwell, 1977). This, according to evolutionary economists such as Pavitt (1991), is crucially related to the development of firm-specific competences and their continued development within the corporate organization.

The most innovative firms, in terms of the appropriation of invention through patenting activity, are seen by Pavitt as generally being the largest and longest lasting. Their superior access to resources often includes an R and D capability that makes it unlikely that they will be surprised and, if so, ensures that they will be well able to overtake new entrants or old competitors with improvements on early product or process designs (Pavitt, 1988). Clarke (1988) has suggested that, since the typical large firm operates across a portfolio of product markets, it is likely to be operating at all four phases of technological/market development at any one time, and is therefore able to spread its risks.

Teece (1977) first put forward the hypothesis that the distinctive competitive advantage of the multinational corporation is its ability to use its tacit competences across a diversity of global market places. The internal ability to adapt to radical changes in products or process design is also seen to depend on the "relatedness" of specific competences. Both Rumelt (1987) and Teece (1977) stress the importance of complementary competences or "assets" in the design, production and marketing of new products. The recent historical analyses of Chandler (1990) and Porter (1990) also lay emphasis on the abilities of successful national or regional communities, in bringing together their competences in a complementary manner. The latter author speaks of the importance of appropriately sophisticated domestic markets, of specialized factors of supply, especially in downstream suppliers, and of well-matched regional structures and technologies.

The required dependence on inter-firm and inter-sectoral networks (isomorphism) does not allow rapid changes. First-movers often fail because of the strength of existing ties, by means of which sectoral markets operate and achieve their separate collective identities. Established relations along the value-chain also appear to exist across sectoral boundaries in a manner which underpins ongoing stocks and flows of information and logistical resources (Pavitt, 1986). Furthermore, the firm exists within a politically regulated environment in which national and supranational governments provide extra-market support. Such institutionalized inertia can provide barriers too great for prevailing oligopolies to be easily challenged from outside the sector.

Yet it is the comparative success of some organizations and some communities in adapting to change that is stressed in both of the latter accounts. In this respect there are many similarities between transformational and evolutionary descriptions of change, in the emphasis placed on the macrosystemic characteristics of successful transformation. It might well also be argued that in concentrating on survivors of the climacteric of the 1970s and 1980s, evolutionary economists have neglected the circumstances surrounding the terminal crises of the non-survivors, and perhaps even those whose survival might be attributed to the educational effect of crises within the firm.

There are few rigorously described accounts of innovation from design to implementation, nor indeed of the critical junctures in the transposition of new paradigms within the firm. The highly political nature of the process seems often to be discounted in economic modelling against the ultimate inevitability of the emergence of an efficient new operational paradigm in the form of "common sense" or routinized procedures represented in a new technological trajectory (Nelson and Winter, 1982). The acquisition of new knowledge therefore appears as a function of time and codification of fresh information — presumably accompanied by attrition among slow learners. Thus the adoption of a new paradigm can be generalized to a shift between two ideal type production functions (or innovation trajectories), in a manner that appears dangerously close to technological determinism.

In most Anglo-American case histories of technological innovation, whether brought about by the adoption of external inventions or through a process of internal design, the process is presented as one of conflict. Both Wilkinson (1983) and Whipp and Clark (1986) suggest that junctures between the stages that mark the passage of design or the adoption of a project towards its eventual implementation and use in routine operations can each become the locus of contestation over the manner in which the next stage should be carried out. The propose-dispose dialogue that takes place represents a segmentation of interests along several dimensions (Kanter, 1983). Case studies have suggested that the most profound conflicts in values and beliefs occur between the R and D and marketing functions. Some observers see the social and spacial location of the former *vis-à-vis* other personnel as crucial to the reduction of impediments to understanding (Woodward, 1965; Francis and Winstanley, 1989).

Modern economists have been inclined to regard such intra-firm conflict as merely frictional. Nelson and Winter (1982) distinguish between three levels of "routine" within the dynamic processes of incremental innovation. These can be readily identified with functional descriptions of three possible levels in organizational hierarchy. In common with descriptions of hierarchy advanced by Williamson and other New Institutional economists, these functions are seen to be carried out in a totally complementary fashion. Williamson (1982) has, however, proposed that the nature of transactions within the internal market of the corporation will reflect in a conscious segmentation of so-called governance structures, in the manner illustrated in Fig. 5.2.

The clan form of organization (Ouchi, 1980) is extended to employees with highly specific skills of an unmonitorable nature, such as senior management, while professionals with generalized knowledge can only be maintained in loose "primitive teams". Those applying firm-specific skills to monitorable tasks must be motivated by permanency of employment and deferred rewards such as pensions which generate an "obligation". Their incentives are contrasted with the spot payments that are all that should be made available for those with monitorable and non-specific skills. (In a similar model put forward by Mok (1975), these latter are described as being in a secondary labour market.)

What is, perhaps, surprising is that in spite of the widespread adoption of Polyani's (1964) description of tacit knowledge as the source of specific competencies that provide the firm with its competitive

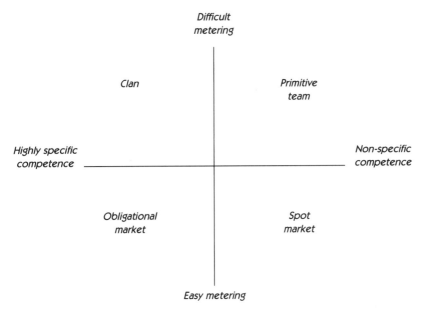

Figure 5.2: Model of internal governance of the firm, adapted from Williamson (1982). See also Mok (1975), Loveridge (1983).

advantage, there has been little attempt within the economic modelling of the internal labour market to provide a polar variable around the intellectual "technic" of the knowledge base underlying professional competencies (Jamous and Peloille, 1970). This might, for example, incorporate the transaction costs of overcoming the impediments to communication and understanding represented by disciplinary boundaries of the kind described above. Since these can be seen to have developed with increased educational and professional specialization within society at large, they might appear non-trivial in the coordination and control of the firm's specific competences (Pavitt, 1991).

One example might be found in the role of "product champion", isolated by Schon (1963). This emerges as a key to successful innovation in the situational model constructed by the Sappho II team (Rothwell *et al.*, 1974). The requirement for sponsorship from among senior management (i.e. within the clan), and of the continued support of a chief executive, suggests that the technical innovator is likely to be relatively isolated from authority (perhaps within a "primitive team"). The mobilization of resources for non-routine activities is seen to require, at the very least, that the sponsor is able to ease proposals through appropriate administrative channels designed to filter out non-routine actions.

A parallel might be drawn here with the key role assigned to the translator by Kuhn (1975) in mediating between incommensurate paradigms in normative science and, more particularly, in explaining the new and revolutionary applications of a familiar (scientific) language. Often the role of product champion can be seen as one of explaining the likely outcomes of a project to technically illiterate executives in a familiar (non-scientific) language. For example, Brady (1984) has suggested that, across a wide variety of cases in the introduction of IT, senior British management did not possess sufficient knowledge to assess the nature and implications of the investments they were making. But, as Metcalfe and Boden suggest elsewhere in this volume, the company strategist will be primarily concerned with matching core competences in configurations that are seen to meet the needs of "the market" in the eyes of that person's most significant stakeholders, or "strategic set" of collaborators and competitors. The recognition and empowerment of actors along this network is an essential part of the ongoing role of the strategist. Much of the manner in which this relational network is defined will depend both on values, beliefs and goals, and on identification of areas of risky dependency or perceived contingent uncertainty in relation to the latter (Loveridge, 1989). In the small science-based enterprises of Silicon Valley these are likely to be different from those facing the strategist in a multinational corporation such as Hewlett Packard.

However, as Offe (1985) puts it, all markets are the product of pre-existing institutions; or to use Casson's (1982) terminology, the entrepreneurial firm conjoins pre-existing social networks. The psychic costs of moving to new markets might therefore be expected to be high. One of the apparent paradoxes contained within the post-Fordist paradigm is its exponents' use of exemplars that appear institutionally entrenched within traditional social networks, such as exist between artisanal enterprises in the regions of Emilia-Romagna, Baden-Württemberg — and of course Japan. The paradox is contained both in the longevity of the collaboration that exists between apparently similar, and therefore *ipso facto* competitive enterprises, and in their prior existence in an epoch that was apparently dominated by the autarkic Fordist model of the firm.

In fact there is very little historical evidence that Fordist modes of production and administration dominated all industrialized manufacture prior to the crisis of the 1980s. Empirical studies such as those of Woodward (1965) and the Aston Group (Hickson *et al.*, 1969) suggested the existence of a range of organizational modes; a taxonomy that might well indicate a configuration of interdependent sectors similar to that put forward by Freeman *et al*. (1982) and Pavitt (1986). If this were so, then organizations in the Fordist mode might well have played a part in shaping the viable forms of related suppliers and distributors by virtue of their monopoly power in mass markets for consumer goods.

However, it remains true that national differences in organizational structure have been distinctive enough for observers such as Child and Francis (1979) and Chandler and Daems (1984) to suggest that only American manufacturers showed a significant trend to the adoption of divisionalized/mass production, while those in other countries tended to remain at earlier stages of the evolutionary trajectory taken in the USA. British firms, for example, tended to remain as loose holding groups employing small or medium batch production modes, while German firms retained functionally specialized hierarchies with highly central-ized general management. In these terms, the British corporation could seem to be much better placed than the German to engage in what Miles and Snow (1986) described as "dynamic networking", or the out-placement of risk in alliances with other firms — a form of strategic behaviour often characterized as part of the post-Fordist paradigm of organization.

THE NATURE OF PARADIGMATIC CHANGE

It might appear then that the term "paradigm" is used to indicate the existence of a cognitive archetype or mental set (though of widely varying dimensions across different accounts). A number of writers on strategic decision-making, working within a social anthropological tradition, have likened a crisis in corporate management to that which is said to accompany revolutions in normative science (Miller and Friesen, 1984; Johnson, 1987). They revert to the concept of a mental paradigm utilized by Kuhn (1962, 1975) to describe a set of logically consistent cognitive elements that are held in common with a community of significant others (in this case the corporate executives).

These elements include "core values" relating to the need for control and prediction, a set of "metaphysical beliefs" about permissible analogies and metaphors, "symbolic generalizations" relating to causa-lity and, most importantly, "concrete exemplars" that demonstrate the multivariate applications of the latter (Kuhn, *op. cit.*, pp. 182–7). For these writers, as for Kuhn, a revolution occurs only after a long period during which experiential "puzzles" defy explanation within currently accepted generalized laws and axioms. This gives rise to alternative explanations which are rejected or are totally incomprehensible to the majority of the community. Crisis occurs when the legitimacy of the prevailing elite is challenged as a result of an event that triggers the overt articulation of a viable position which challenges the prevailing hegemony within the organization. Such positions fall within the category of those described by Rumelt (1987) as "framing theories". Their proponents sometimes attribute powers of bounded rationality to the strategic decision maker (Simon, 1957), including the rationally instrumental use of strategic sets of other corporate actors as referents and the use of a portfolio of behavioural "structural repertoires" (Clark and Staunton, 1989).

Other theorists believe decision-making environments to be constantly "enacted" through the application and unconscious adaptation of habituated "recipes" (Weick, 1987; Spender, 1980). The latter term has found a wider currency among teachers of management to indicate the existence of a shared, stable, and consciously adopted set of comparators that make up a mature sector, together with recognized modes of behaviour within that sector — in fact a strategic paradigm (Grant, 1991).

The present author has adopted the more traditional position in sociology and social psychology of distinguishing between the normative frame of reference which has been internalized by the decision-maker during an ongoing process of socialization, and the more instrumental and, even, habituated responses made in their everyday roles (Loveridge, 1989). This has been described as the frame-and-formulae approach (Johnson, 1987). Following earlier notions of reference group theory (Merton and Kitt, 1950) the author sees the strategic frame as comprising somewhat abstract exemplary archetypes, such as the Fordist philosophy (Ford, 1922) on one hand, together with more immediate and concrete comparators within the membership group of the industry on the other; say the multitude of jobbing plants making up the West Midlands engineering industry (Bott, 1954).

Similarly, strategic actions comprise abstract mission statements and the projection of visionary goals through a variety of media, primarily nowadays the annual report to shareholders. These may be regarded as formal statements of what Argyris and Schon (1978) describe as "expounded ideology". In the routines of day-to-day behaviour, evidence of deeper and more immediately reinforced "ideology in use" may be found. In the manner illustrated in Fig. 5.3, this behavioural pattern, or formula, may well relate to deeply embedded archetypes, while publicly espoused mission statements are more likely to be shaped by what are seen to be the needs of perceived stakeholders or comparators in the same business.

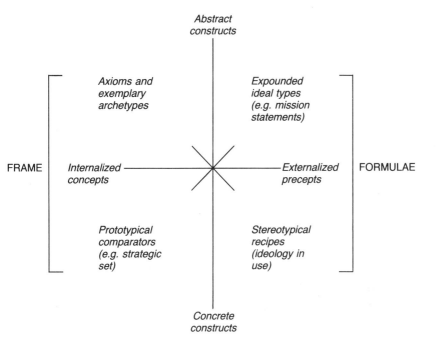

Figure 5.3: Suggested elements in the cognitive paradigm and possible direction of cognitive shaping of activity (Loveridge, 1989).

Uncomprehended change can lead to consternation, more especially in the former cases through the resultant cognitive dissonance. Puzzles that defy deeply embedded personal recipes imply a high cost in shifting referents and/or axioms. Threats to internalized values and archetypes, accompanied by lack of morally normative guidance, are likely to induce anxiety and emotional responses. Mannheim (1944) described this state as a "paradigmatic experience", a term Gouldner (1954) used in describing the trauma experienced by managers during the first labour walk-out in a remote rural gypsum mine in USA.

This second description of paradigmatic change is not necessarily one to be applied to the experiences of all strategists in a crisis situation. The strategic frame used by senior executives is unlikely to reflect the holistic consistency of the "objectified" structure of an intellectual discipline. Neither are archetypes nor experiential prototypes likely to be as central to the ego-identity of the business executive as to the academic researcher or scientist. Commitment to the espoused business mission may be totally "rational-instrumental", and largely designed to provide an umbrella of objectives that bring a concurrence in action across a range of incommensurate explanations of corporate "puzzles". These latter usually reflect the separate values and logics of actions within professional and functional disciplines present within a modern multinational corporation (Karpik, 1972). The achievement of satisfactory outcomes for the disparate groups of stakeholders making up the enterprise might be said to be a necessary condition for the survival of an institution. But the preferences of such stakeholders are not necessarily free from influence by its leaders. If Kuhn can be said to have demonstrated empirically that a collective leadership exists within the scientific community, a wealth of prescriptive management literature conveys the perceived necessity for leadership in modern corporate decision-making to the would-be business executive. Furthermore, anthropologists and sociologists, most prominently Foucault (1973), have demonstrated that the use of language in the everyday discourse of organization carries an implication of hierarchy through a wide variety of firm-specific and nationally-specific cultural styles. Boundaries are drawn and claims to legitimate authority are often made in the affectively neutral and depersonalized modalities of procedural techniques, such as those of accountancy. These are used to justify executive action even where, as in decision making on entirely novel, open-ended commitments to investments in R and D, their use can be no more than symbolic (Freeman, *et al.*, 1982). Broader in function, the use of inspirational rhetoric in the form of corporate mission statements of beliefs and objectives has become a widely adopted mode of response to the experience of normative uncertainty among strategic leaders.

CULTURE AND CONTINUITY

For the researcher, these different levels of action orientation pose evident problems of distinguishing between the espoused ideals and mission statements of managers and their operational stances and orientations. For example, the elevation of JIT/TQM systems of control over corporate and intercorporate relations to that of an ideal type in the business and academic literature adds authority to its effect on managerial behaviour. But reports on the operational behaviour of both executives and their subordinates in the West are somewhat mixed.

Although most major manufacturers of component sub-assemblies in the West now claim to be orchestrators of their design, it is rare to discover a customer among major automobile manufacturers who will admit to this degree of recognition for any one of their so-called "preferred suppliers". In two major surveys of component suppliers largely located in the Michigan/Ohio area of the USA, Helper (1990) discovered attitudes of profound mistrust between suppliers and customers throughout the mid and late-1980s. This was largely due to the belief among the former that, in the mode of operations prevailing in North America, JIT entailed a simple down-loading of costs by their customers, with none of the assurance of continuation and collaboration in the supply relationship that characterized operations in Japan. Much the same distrust was expressed by vehicle component suppliers in the West Midlands who took part in a much smaller sample taken by Turnbull (1986). What is interesting is the implication behind statements made in both surveys that, at each level of the value chain, the informant would be willing to collaborate if only his or her customer would also change.

For the evolutionary economist, such phenomena might be described as external learning costs. These could ultimately be debilitating for the whole system, defined, in this case, to take in the entire nationally or regionally located value chain. No consideration is usually given to the "core values" and set of "metaphysical beliefs", to use Kuhn's terminology, that underlie particular "ideal types" and "concrete exemplars" in use within those culturally as well as geographically defined systems.

One approach to this form of analysis is that of analysing the unfolding of strategic responses emanating from a chosen firm within the historical context of sectoral or regional change (Whipp and Clark, 1986; Smith *et al.*, 1990). Within this process these writers identify core beliefs or ideologies discernible in the reiterative enactment of what Clark and Staunton (1989) prefer to describe as being "repertoires of plays by theatre groups and footballers...", not reproducible in a simplistic manner. Their reproduction is both precarious and has a tendency towards what Giddens (1985) has termed "chronic recursiveness" (Clark and Staunton, 1989, p. 29). Breaking this mould is, evidently, both a political as well as an intellectual learning process, involving the multiple transactional strands bought together at the strategic conjunct markets and hierarchies.

In the following section the histories of two companies, Lucas Industries plc, founded and based in the English West Midlands, and Robert Bosch GmbH, founded and based in Baden Württemberg, are compared. Over both life spans revolutionary changes in technology have been accompanied by severe threats to relations within the supply chains to the automotive, aerospace and telecommunications sectors in which, to greater or lesser extents, both companies have operated. They were, and are, both threatened by what has earlier been described as architectural change, involving the disappearance of old customers, suppliers and competitors, and the emergence of new, very differently organized comparators. It is hoped to demonstrate the resilience of a pattern of action in each company that has its roots in events surrounding the foundation and adaptation of corporate forms in their respective businesses. The evidence has been drawn from official histories, archival material and interviews and observations carried out by the author. In the case of Lucas these began in 1976, in Bosch in 1984.

CONTRASTING MODES OF LEARNING

The date of their formation — (Joseph) Lucas Industries in 1875 and Robert Bosch GmbH in 1886 — enables the evolution of both companies to be traced in relation to three possible long cycles of economic activity (Nockolds, 1976; Loveridge, 1990). Their choice as cases in innovation has been determined by the position they occupy as subassemblers or first-line suppliers to a range of complex machines regarded as central to the development of mature consumer economies. Both firms manufacture electronic and electrical equipment essential for engine, body and braking control in motor vehicles. Both supply the telecommunications and domestic machinery sectors. While Bosch is a major provider of final products in these latter sectors, Lucas now has a marginal and decreasing interest. By contrast Lucas has expanded considerably its capability for the design and supply of aerospace components over the last quarter century, while Bosch (who invented the first fuel-injection system for the Mercedes Benz engine of the Messerschmitt 109) has largely withdrawn from this market.

In terms of most indicators of intellectual and material success, Bosch has out-performed the British company for much of this century, apart from the immediate post-World War periods. Both companies are currently among their national leaders in patenting activity, Bosch being more active than Ford Motors in the USA, and only overshadowed by the major Japanese combines and General Motors (Patel and Pavitt, 1988). The annual sales of Bosch are greater by a factor of three than those of Lucas, fixed assets of the former are nearly four times greater, and the number of employees is nearly three times greater (1988 Report and Accounts). Bosch ranks as one of the largest engineering manufacturers in the world, and is Europe's leading automotive component suppliers by a large margin (Slade and Fordham, 1990). However, both companies are relatively unique in remaining independent of the large holding groups that now direct the long-term fortunes of all of their remaining competitors. (This has been due to financial strategies which, though crucial to a wider understanding of their strategic paradigms, has to be largely ignored for the sake of brevity.) Lucas might be said to play an important global role as an alternative supplier to in-house contractors, as well as being a preferred supplier across a range of marques in the automobile and aerospace sectors of Europe and the USA.

It is apparent that the decline in their markets, particularly the automobile market, hit both companies. Both managements reduced their labour forces over the early years of the last decade by between 30 per cent and 40 per cent. But whereas the value of Bosch sales continued to climb by around 6 per cent p.a., in 1981 Lucas made its first-ever loss — an experience described as traumatic by then senior executives (Van de Vliet, 1986) and only slowly climbed back to profit.

Over much of the preceding century of their existence, both companies had adopted highly centralized administrative structures. Both had been greatly affected by the leadership styles of their founders, in the case of Bosch by the longevity of the latter's personal influence, up to his death in 1941, and in the Lucas Group by three generations of family chief executives up to 1948. Both were pioneers of flow-line production, and had built model factories before Henry Ford established his first workshop. They enthusiastically identified with the latter's world philosophy and methods in the 1920s and 30s. In two World Wars and their aftermath, Lucas were employed by the British government to propagate scientific management among other British manufacturers. In the 1950s both groups began to move their diverse activities into divisionalized structures. In the case of Lucas these changes generally followed the succession of each of the three former personal assistants to Oliver Lucas, the last of the family chief executives. In the case of Bosch the centralized administrative structure has remained much the same between 1963 and the time of writing (1991). From the former date until 1984 the same chief executive officer remained at the helm, and still occupies an influential role on the supervisory board. The present CEO is only the fourth since the resignation of the founder in the 1930s.

Lucas's response to crisis was as dramatic as the circumstances surrounding it. It coincided with the election of the previous MD to chairman, and the appointment of two successors to his former job.

After an abortive attempt to relocate and separate the activities of the automotive electricals division, one of the new appointees resigned. In 1983 under the title of Competitive Achievement Plan, the Board devolved all operational control to its business units, subject to central appraisal and evaluation of performance against bench-marks set in relation to the best of their immediate competitors. Having cascaded their trauma, the Board then provided the potential escape to a new "normalcy" through the agency of newly created Lucas Engineering and Systems Limited (LES) as an internal consultancy unit. Under the leadership of Dr John Parnaby, a former Professor of Production Engineering, LES offered a new organizational template based on the best principles of Japanese product engineering. The blueprint was that of JIT cellular production, in which work is "pulled" through a value chain that extends beyond the factory walls to suppliers and customers. What is notable about the Lucas system is its conscious dependence on labour flexibility, rather than on automated systems of an FMS variety. (One notable exception was an investment in a fully computer-integrated plant at Telford, which closed after a few years of operation.) Since that time Lucas has become known as one of British manufacturing's prototypes within the new paradigm of flexible specialization.

By contrast, Bosch began its transition to flexible manufacture in the 1970s, largely based on its core capability in the design of machine tools. The company's reduction in labour has largely been achieved without plant closure, but by the automation of existing lines. The production of machine tools has actually expanded as a proportion of Bosch Group business over the last decade. Over this period Lucas has been withdrawing from the manufacture of hardware in favour of software design and specialization in areas involving sensing and actuator technology. Bosch movement into FMS does not appear to have been the result of a battle between "product champions", as at Lucas, but rather the outcome of opinions expressed at the monthly central meeting in Stuttgart between all managers (MDs) of strategic business units.

Executives in the two companies are very aware of the other's market strategies, especially in automotive components, where they are direct competitors. Many organizational and educational initiatives in Lucas seem geared to Bosch's example. But formal similarities disguise profound differences in underlying behaviours and beliefs. The "Lucas System" of scientific management was an ideology developed against a background of the fragmented "jobbing" structure of West Midlands engineering (Loveridge, 1982). It seems doubtful whether it ever operated in the manner presented in the texts written by Lucas engineers. To variability in supply has to be added variability in product design created by the individualist demands of the several British automobile assemblers, and the need to cater for a large after sales (replacement) market. Both suppliers and customers greatly resented attempts by "Uncle Joe" to enforce standardization. Day-to-day relations across companies and across all stages of production within plants were, by the late 1960s, marked by persistent conflict and recrimination.

The role of line foremen became a focus of this conflict since, deprived of technical authority, they became agents of the numerous specialist work study, production engineering and progress-chasing managements, with little brief other than unrealistic numerical performance targets. A second form of conflict, to which we shall return, was that of the status of the "companies" making up the Lucas Group. Much of the Group's growth had been by acquisition over the 1920s and 1930s, and for both operational and financial reasons the companies thus acquired were, up to a 1961 decision by the Monopolies Commission, allowed to retain their separate trading identities. After that date, growing intervention by Group (housed in the old Lucas model factory) provided another source of conflict. A major expansion in 1968 added to these centrifugal forces when, on the advice of Government, the Group acquired most of the existing British suppliers of aerospace fuel supply, ignition and body control subsystems.

By contrast most of the Bosch growth has stemmed from the organic development of products and technologies in-house. Strategic alliances made with telecommunications companies in the 1930s provided the Group with access to new markets. Small stakes in joint ventures were increased in a gradual takeover, over which time relations of mutuality have been developed. However, much of Bosch's success has been attributed to its embeddedness within a regional industrial order held to exist in its birthplace of Baden-Württemberg. Central to this explanation is the existence of multiple small subcontractors specializing in one aspect of the value-chain leading to the assembly of complex machines (Herrigel, 1990). In this respect the region may be considered not dissimilar to the West Midlands, nor indeed to the Michigan/Ohio context that also gave birth to the automotive industry. Unlike the latter, these small firms are seen to have been extraordinarily successful in the application of new technologies, and in adapting to the requirements of the more demanding market created by Japan and by global competition.

The key to their success is seen as being their willingness to collaborate with local competitors in the creation of an infrastructure that has, to some extent at least, "socialized" the risks of change. Trade associations have long played a proactive role in Germany in regard to technology transfer and in providing an ongoing source of information, and research backup in regard to new markets and new technologies. Precompetitive product development is orchestrated through local colleges and universities. Local banks have also been involved in such ventures and have not simply acted as linesmen in judging the play of industrial activity in the region. The regional government has been active over the last two decades in providing direct subsidies for technological development, as well as coordinating many activities through legislative bodies and a large semi-private foundation. Perhaps most important has been the collaboration between schools, colleges and universities in providing both vocational education that complements the apprentice system in all sizes of company, as well as in providing research and development facilities. The graduates of these programmes provide an alumini network across contracting firms and regional industries.

There is, indeed, a remarkable match between the qualifications provided by the vocational education system and the hierarchical grading within German companies. Unlike the movement toward multi-skilled crafts and cellular-based operatives in Lucas and other British manufacturing plants, Bosch has retained specialized crafts operating in traditionally centralized planned maintenance units. Operatives are, however, expected to attain higher off-job educational qualifications and to undertake a range of everyday maintenance tasks on automated lines. The "Meister", or shop foreman, continues to act as a fulcrum in maintaining technical and social balance within the hierarchy. The role, which has been crucial to technological innovation in the past histories of both Lucas and Bosch, is now increasingly occupied by university graduates in both companies.

Effects of environment on invention–innovation

The role of these companies as carriers of new technology and as multiplier/accelerator mechanisms in its diffusion has been largely discounted in many empirical assessments of national competitiveness. Yet it is apparent that most improvements in both land and aerospace vehicles has derived from subassemblers (Abernathy *et al.*, 1983). Robert Bosch built his company around his invention of the low-voltage magneto which allowed the development of practicable internal combustion engines. Subsequent expansions in markets were the result of in-house inventions or, as mentioned earlier, inventive synergies created through strategic alliances. The Lucas's were self-taught engineers who developed a facility for adapting and patenting relatively low-tech component products and producing them cheaply in large batches. The accountancy-trained CEOs who were their immediate successors retained this view of their business. Sir Bernard Scott was reported as laconically remarking to the official historian, "I don't think Lucas ever invented anything much" (Nockolds, 1976, p. 403). On first reading, the statement is baffling. Not only had the company retained a central development unit since the 1900s, but since the early 1950s the research function had been represented on the Board. Moreover, after initially purchasing the Bell Telephone license to produce semiconductors, the company had succeeded in inventing the first viable alternator, that was to replace the magneto in car and truck engines. In its way, the development can be likened to Bosch's original breakthrough. It was licensed to the General Motors subsidiary Delco in 1964. Owing to the conservatism of its British customers Lucas itself did not have the opportunity to fit the alternator to production models until 1971, by which time Bosch was already supplying Mercedes Benz with its own design.

Throughout the 1960s Lucas held its place as one of the six largest producers of semiconductors in Europe (Sciberras, 1977). What is evident from all contemporary accounts is that the development of the technology was localized, and heavily reliant on tacit knowledge (for example, Langrish et al. 1972, p. 344). Although attempts were made by some operating companies to incorporate semiconductors into new products, little headway was made in marketing them, and, especially in the closed competition for Post Office telecommunications, the financial and knowledge investments made by the Group appear to have been totally insufficient. Even the large internal market was under-exploited. Partly this was because of an apparent incomprehension on the part of production and mechanical engineers of the material processes involved, and of the need for changes in attendant operating procedures. There is also evidence that some divisional managements saw their incorporation into existing designs as undermining existing markets, particularly their replacement market. The later development of mass producers of chips and boards to whom Lucas continued to sell licenses up to the mid-1980s, reduced internal development to that of an in-house alternative supplier and customizer for the development of devices by other business units in and outside the Group.

It might well be said that R and D held a high symbolic value for Lucas over much of this century. Individual research directors and project managers figure as prophetic heroes in company myths of "what might have been". They had little significant impact on "making brass", however. This was partly to do with a strategic concern with achievable short-term production and financial goals, and partly with lack of central control over the processes of operational implementation. In the context of such a fragmented environment, it appeared wholly apposite for the Board to adopt the strategy of pushing R and D down to the plant level over the early 1980s, as part of a move toward the so-called "simultaneous engineering" of new products. This is also much closer to the reality of incremental improvement in product and process design achieved at plant level in the past — as opposed to the Group's history of relatively unexploited achievements of central R and D.

These activities are now carried out at Divisional level and are largely designed to service plant needs. (See Metcalf and Boden in this volume.)

Again the contrast with practice at Bosch is remarkable. Research in the German company is a largely group-financed activity (90 per cent), located next to the headquarters building. It has generally employed something over ten times as many "scientific" and "developmental" staff as has Lucas Group (central) research. These people appear to work in relative isolation from production units, even though up to 80 per cent of the work done is said to have been generated upwards from the plants. The Board has recently announced that it intends to deploy older R and D staff to production units, retaining only younger people at the head laboratory for work of a more "blue sky" variety.

A major bridge was created through the conscious development of corporate sales teams in the 1960s, whose job was and is to liaise with Bosch's largest ongoing customer-users in the development of existing and proposed new products. These four-person teams have to negotiate with their sub-divisions and then, directly, with divisional management for an R and D budgetary allocation for this purpose. Needless to say, the sales team are generally doctoral engineers. They provide an important conduit between the central R and D, the customer and the plant engineer and the Meister.

However, there are direct links between the centrally employed design and scientific staff and those of corporate customers, which appears to have resulted in close collaboration between Bosch and Mercedes Benz in the prototyping of early applications of semiconductors. This was done through their incorporation in prototype engine designs from the early 1960s (the contrast with the ill-fated BRM racing car project on which Lucas tested its devices might be noted in passing). Bosch continues to produce chips in its K8 sub-division of the Automobile Equipment Division, as well as specialized devices in the Telecommunications Division. Like Lucas, it buys in most standard devices but regards the K8 plants as offering a potential competitive advantage in working with its customer-user in customized fields.

At the 1990 Annual Meeting of Lucas plc the chief executive declared the Group to be committed to a policy of organic development through an increased expenditure in R and D, in spite of the recent fall in profits. (There has, in fact, rarely been a time when Lucas has not invested a greater proportion of sales income in R and D and employee training than most British manufacturing companies.) At present its focus, as already mentioned, is on the improvement of present products together with the development of systems engineering software and sensors (having otherwise withdrawn from industrial tooling). Towards this end the Group has acquired a number of American-owned companies through significantly increased institutional borrowing. It has also strengthened both its capability and position in the design chain servicing the major US aircraft and defence suppliers through acquisitions in North America. Both Bosch and Lucas continue to enter alliances with Japanese and other European manufacturers, involving either the exchange of licenses or joint-ventures. Recently Bosch acquired the telecommunications division of the failed AEG giant in a government-inspired rescue package mounted with Siemens. However, such coordinated action at the strategic level is becoming increasingly difficult as other large German manufacturers compete with Bosch for the exclusive design of integrated telecommunications and vehicle products. Finance houses such as Deutsche Bank and Allianz Insurance, which had become the orchestrators of strategic change in post-war German industry, appear increasingly to have turned their attentions to the problem of their own survival in a globally competitive market for financial credit. With the exception of power tools, Bosch's consumer products face potentially overwhelming Japanese-owned competition. It would be difficult to suggest that crises are entirely past history for either company, or for their product markets.

DISCUSSION

Differences between the interpretation and enactment of what has been described as the "post-Fordist" paradigm are clearly apparent in the histories of these two companies. Much of this difference must be attributed to the respective product development and design capabilities of the two organizations. These characteristics have to be interpreted in somewhat wider terms than the firm-specific competences described by

Pavitt (1991) to take in the ability to invest and/or translate developments in technological ideas into effective products and processes (see
Whipp and Clark, 1986). Any assessment of this capability must, then,
involve some consideration of the "non-trivial" costs of bringing about
value added collaboration between possessors of specific competencies,
particularly where these are based on professional or scientific bodies of
knowledge that have been framed and codified in a manner that entails
high entry costs to proficient task performance (part of the "technicity"
referred to earlier).

A second evident characteristic of the history of the two companies is
the recursiveness of their strategic recipes. The post-Fordist Lucas
management continues to acquire new competences and to enter new
markets through appropriation, in much the manner of their entrepreneurial predecessors of the late 1920s and 30s — whilst emphasizing a
commitment to organic growth. Bosch continues to invest in centralized,
long-term R and D, even when quite clearly beset on every side by
challenges from new entrants and old collaborators (like Siemens and
Mercedes Benz) across market boundaries that were previously agreed
and maintained by financial stakeholders. In the day-to-day discourse of
both companies, the engineering metaphor of "total systems" approach
to organizational problems has shaped executive responses for over a
century.

Yet the maintenance of the administrative and occupational hierarchy
within the German Group must be contrasted with the radical
restructuring of Lucas between 1982 and 1987 (when there was another
change of CEO, heralding further rationalization in the Divisional
structure). On the basis of the brief account of the strategic environment
of each firm that has been presented here, the reader might well fall back
on a straightforward explanation of a matching of structures to external
contingencies: this being delayed in Lucas by the persistence of the
Fordist archetype. To do so would leave unanswered the question of
how hierarchical control exercised within the Bosch hierarchy has come
to match so exactly that created within the broader operational context
of the German, and more particularly, the Baden–Württemberg "industrial order" (Herrigel, 1990). By the same token, one might well require
some explanation for the persistence of Lucas's executives in seeking to
impose their ideals on an internal and external environment that was
antithetical to their most basic precepts.

In this connection, Williamson's (1982) typology of the internally
segmented governance structure of the firm (see Fig. 5.2) has, of course,
been extended by him and by others to classify forms of inter-corporate
relationships (Boissot, 1982; Teece, 1987). There is little in the reasoning

underlying this model to indicate the basis for the trust relations indicated by the existence of clans within the context of the individualistically opportunist market place that it presupposes. For the most part, explanations are presented in terms of the mutuality of shared risk through investment in a joint project (whether a strategic alliance or a professional career). Little account is taken of how the parties arrived at a realization of the benefits to be derived from this mutual acceptance of risk; indeed the explanation appears somewhat tautological!

Brusco (1990) uses a similar explanation of how the communal provision of what are described as "real services" (external economies) helps to create industrial districts such as Baden–Württemberg. However, he supplements his explanation with a description of the fusion of commitment brought about by the local coalitions of Roman Catholicism and Eurocommunism in the formation of the artisanal culture of the Emilia Romagna region. In other words, the "exogenous" explanation lies in pre-existing institutional forms and in the manner in which entrepreneurs conjoined them to create a market place (Offe, 1985; Casson, 1982). In the German case, as in the Japanese, the remarkable degree of continuity brought about by the continued dominance of military-landowning elites in the industrialization of their respective countries must be regarded as significant in maintaining an inter-corporate hierarchy (Loveridge, 1983). To this explanation Herrigel (1990) adds the independent influences of early collectives of small artisanal enterprises which, like those in other parts of Northern Europe, respond to British and American technological innovation in a proactive rather than defensive fashion.

But it also seems important to include the role of innovating agents such as the inventor–engineer Robert Bosch. In 1913, when his firm had already established an economically significant position in the nascent automobile and aircraft component markets throughout the world as well as in the Württemberg region, his organization took an all-out strike. It was a time of political unrest throughout Germany, as craftspeople subsumed their industrial goals beneath those of the Marxist SDP. Bosch responded by joining the local trade and employer associations, in order to persuade them to follow him in financing vocational education and training across the region on a scale that seems difficult to contemplate from a British standpoint. By contrast the British Engineering Employers Federation remained locked in a struggle with craft unions, including the newly formed Electrical Trades Union, over control of the apprenticeship system. This went on until the system was largely abandoned in the 1960s, to return in a vastly modified form over the last decade. The recognition accorded to technical and professional qualifications, although not so overtly confrontational, has been reluctantly tardy. Lucas stand out among British engineering employers for their contribution to vocational education (and more recently to staff development). It retains much of the internal dichotomy between administrative (managerial) competences and those of technical designer–manufacturer observed to exist in British companies by Sorge and Warner (1986), though not in German companies. However, the Group only reluctantly recognized trade unions, and has not hesitated to adopt aggressively confrontational stances in defence of managerial prerogatives. It could not, in any case, easily escape from its context in Anglo-Saxon market relations. Thus the transactional boundaries between markets, corporate and artisanal, can be seen to have been forged through the manner in which new skills and their underpinning knowledge disciplines were sponsored and recognized by client–customers during periods of technological transition. In the case of Bosch, his example provided the basis for establishing a relationship between the internal hierarchy of the company and its occupational context that proved mutually reinforcing in generating commitment to both. As such, his actions were not atypical of those of other large German employers working within an institutional frame provided by a long history of enlightened sponsorship by Länder princes (Herrigal, 1990).

In this manner the complementarity between formal education and career in any part of the interdependent enterprises making up the business/technological system is made apparent to the new entrant at an early stage. Much the same process may be observed in the manner in which the Group has incorporated independent partners in joint ventures which have been eventually appropriate. In Durkheimian terms the relationship of the Bosch hierarchy to its relational context might, therefore, be described as being closer to one of organic solidarity than one of a mechanistically imposed division of labour. The existence of this more consensual basis of control might be expected to result in the espoused ideology of the Group's leaders to more accurately reflect their operational recipes than would be the case in a more pluralistic political context.

By contrast, Lucas executives were in an extremely cross-pressurized situation. The company held aloof from local and national employer associations, preferring to see itself as controlling its own internal and external markets. Executives tended to see their company as belonging to a global strategic set of electronic component manufacturers consisting, until the late 1970s, of firms such as Bosch, Bendix, Delco, Hoskiss and Thompson, and more recently of Japanese competitors such as

Nippon Denso. Its executives toured these companies annually, bringing back suggestions for changes in both product and process. Given their recognition by the British government as proselytizers of scientific management, it should not be surprising that Lucas executives saw their company as a conduit for best practice in "total systems management". Unfortunately as described earlier, their efforts were often rejected and resented, both by assembler customers, who preferred to maintain a policy of dual-sourcing among multiple jobbing subcontractors, and by the several large holding groups such as GKN, IMI, TI etc. who retained the Midlands engineering industry in this fragmented structure. Lucas itself was revealed publicly as maintaining a dual policy in this regard, by the 1961 Monopolies Commission report. A number of cross-directorships with the other West Midlands suppliers almost certainly tempered ideals with pragmatism in action.

Both Groups have developed their technological capabilities within the context of the institutional systems which make up their respective national and regional cultures. Keasing (1974) described the former as allowing greater or lesser amounts of adaptation to the degree to which they shaped market or transactional forces in a direct manner. Indirectly, and often with a "culture lag", they contribute to socializing actors in ideas, values, shared symbols and meanings which are transmitted intergenerationally and which can be expected to mould strategic paradigms in a manner that can outlast any particular technological epoch or cycle (Ogburn, 1922). More explicitly political sociologists such as Lodge and Vogel (1987) have advanced typologies based on contrasting clusters of values and beliefs constituting either individualistic or communitarian ideologies. In some later developing countries, historical experiences, including comparisons with earlier industrializing nations, have brought a greater awareness of holism and interdependence at all levels of society, and acceptance of hierarchy based on a similar recognition of communal responsibility in government and business. In countries affected by the Anglo-Saxon values of individualism, the utilitarian view of society as the sum of its parts is made explicit in a number of ways. Most important is that of creating boundaries around specific "properties" owned by individuals or factions, and transacting or contracting for their value with others in society.

These elemental beliefs are, of course, fundamental to most of the frames of analysis used in the discussion of innovation and technology in this chapter. What remains insoluble within this framework are the obstacles to the creation of the "atmosphere" spoken of by Williamson (1985) in which trustful communication and collaboration can take place.

This remains for Lucas as big a problem as it once was within the context of the multi-tiered bureaucracy that its executives attempted to impose on the jobbing structure of British engineering. While adopting a very different ideal of corporate and inter-corporate relationships, its "total systems" operational formula appears often to be felt to be mechanistic and unsympathetic to the individualistic pragmatism that continues to prevail in the context of the sectors of Anglo–American manufacture in which it does much of its transacting.

CONCLUSIONS

In this chapter a number of approaches to technological innovation which have adopted the linear form of the life cycle or trajectory have been explored and critiqued. In particular, the notion of periodic discontinuities resulting in crises of control within individual firms and sectors was found to be subject to numerous situational contingencies. The discussion moved to focus on the cognitive disturbance resulting from a revolution in strategic paradigms of the kind described by Kuhn (1975) as occurring within the scientific community at regular intervals. In the manner of Metcalfe and Boden it was suggested that this paradigm must necessarily incorporate modes of achieving the firm's objectives through gaining and exercising control over both internal (hierarchical) relations and external (market) relations. The need for social control could be seen as particularly significant for strategic management. Their use of expounded ideals and goals might well be instrumental in nature, not necessarily coinciding with operational recipes put into practice. Hence paradigmatic "crises" may not be as profound a learning experience as they appear in public enactments.

This was illustrated through comparisons of historical modes of adaptation in two firms, one British and one German, particularly over the transition from Fordist to post-Fordist modes of organization during the last twenty years. It was found that neither had adopted the Fordist paradigm in all of its elements, but had adapted these to the institutional and ideational context in which their founding company was based. In the case of Bosch this adaptation had proved remarkably viable and had provided the hierarchical/artisanal juxtaposition as the basis of the introduction of a high level of automated FMS. In Lucas, organizational forms went through a profound transformation after an apparently emotional trauma at Board level. However the narrative demonstrated that underlying behaviours and beliefs both before and after the formal devolution of operational responsibility within the company were not as affected as might be assumed from changes in formal procedure.

In practice, both companies displayed a remarkable continuity in many aspects of their strategic behaviour, particularly in regard to the development of technologies and their exploitation. It was suggested that an analysis of the causes of this cultural linearity might well be seen as essential to an understanding of the nature of relative technological learning. Indeed it was precisely because of the erection of culturally prescribed social barriers to the transfer of knowledge that diffusion was not allowed to take place. While this use of the notion of "specificity of assets" has become central to an economic analysis of strategic behaviour at the corporate level, very little analytical attention has been devoted to the problem of the coordination of such specific contributions to the value chain.

In this chapter the manner in which human assets, in the shape of scarce and idiosyncratic competences, are sponsored and recognized (institutionalized) at points of architectural changes has been seen as vital for the future viability of the corporate hierarchies thus formed. Since these changes are often accompanied by the most profound shifts in organizational norms, the role of a visionary agent might become of significance in the reformulation and psychological internalization of new beliefs. Some caution must be displayed in advancing universalistic models of roles such as that of "product champion", however, as the comparisons available in these case studies demonstrate their likely cultural specificity.

References

Abernathy, W.J., Clark, K.B. and Kantow, A.M. (1983) *Industrial Renaissance*, Basic Books, New York.

Aglietta, M. (1975, 1979 edition), *A Theory of Capitalist Regulation*, New Left Books, London.

Argyris, C., and Schon, D.A. (1978) *Organizational Learning: A Theory of Action Perspective*, Addison Wesley, Reading, MA.

Amin, A. and Robins, K. (1990) The re-emergence of regional economies? The mythical geography of flexible-accumulation, *Society and Space*, **8**, 7–34.

Best, H.M. (1990) *The New Competition — Institutions of Industrial Restructuring*, Polity Press, Oxford.

Boisot, M. (1983) Convergence re-visited: the codification and diffusion of knowledge in a British and Japanese firm, *Journal of Management Studies*, **1**, 159–190.

Bott, E. (1954) The Concept of Class as Reference Group, *Human Relations* **7**(3).

Brady, T. (1984) *New Technology and Skills in British Industry*, Manpower Services Commission, Sheffield.

Brusco, S. (1990) "The idea of the industrial district: its genesis", in *Industrial Districts and Inter-Firm Co-operation in Italy*, (Eds F. Pyke, G. Becattini, and W. Sengenberger,) International Institute for Labour Studies, Geneva.

Casson, M.C. (1982) *The Entrepreneur: an economic theory*, Martin Robinson, Oxford.

Chandler, A.D. (1977) *The Visible Hand: the Managerial Revolution in America*, Bellkap Cambridge, MA.

Chandler, A.D. (1990) *Scale and Scope — the dynamics of Industrial Capitalism*, Belkap Press, Cambridge, MA.

Chandler, A.D., and Daems, A. (eds) (1984) *Managerial Hierarchies*, Harvard University Press, Cambridge, MA.

Child, J. and Francis, A. (1979) Strategy Formulation as a Structural Process, *International Studies of Management and Organisation*, Summer, 163–69.

Child, J. and Loveridge, R. (1981) "Capital formation and job creation within the firm in the UK — a review of the literature and suggested lines of research" in *Relations between Technology, Capital and Labour*, (Eds O. Diettrich and J. Morley), Commission of the European Communities, Brussels.

Child, J. and Loveridge, R. (1990) *Information Technology in Europe: towards a micro-electronic future*, Basil Blackwell, Oxford.

Child, J. and Smith, C. (1987) The Context and Process of Organizational Transformation — Cadbury Limited in its Sector, *Journal of Management Studies*, **24**, 6 November. Edited version in Loveridge, R. and Pitt, M. (eds) *The Strategic Management of Technological Innovation*, Wiley & Sons, Chichester.

Clark, P. and Staunton, N. (1989) *Innovation in Technology and Organization*, Routledge, London.

Clarke, K.J. (1988) "Technological Change and Strategic Management", British Academy of Management Conference, Cardiff Business School, Wales, September.

Cotgrove, S. and Box, S. (1970) *Science, Industry and Society*, Allen & Unwin, London.

Dosi, G. (1982) Technical Paradigms and Technological Trajectories — A Suggested Interpretation of the Determinants and Directions of Technical Change, *Research Policy*, **11**(3).

Durkheim, E. (1893: 1964 edn.) *The Division of Labour in Society*, Macmillan, London.

Dyas, G.P. and Thanheiser, H.T. (1976) *The Emerging European Enterprise*, Macmillan, London.

Edwards, R. (1979) *Contested Terrains*, Basic Books, New York.

Ford, H. (1922: 1924 edn.) *My Life and Work*, Heinemann, London.

Foucault, M. (1973) *The order of things: an archeology of the Human Sciences*, Vintage Books, New York.

Francis, A. and Winstanley, D. (1990) "The Organization and Management of Engineering Design in the UK", Symposium on Design for Manufacture, Penn State University USA, August.

Freeman, C. (1979) *The Economics of Industrial Innovation*, Pinter, London.

Freeman, C. (1987) *Technology, policy and economic performance*, Pinter, London.

Freeman, C. and Barley, S.R. (1990) "The Strategic Analysis of Inter-Organizational Relations in Biotechnology", in *The Strategic Management of Technological Innovation* (Eds R. Loveridge and M. Pitt), John Wiley, Chichester.

Freeman, C., Clark, J. and Soete, L. (1982) *Unemployment and Technical Innovation: a study of long waves and economic development*, Frances Pinter, London.

Georghiou, L., Metcalfe, J.S., Gibbons, M., Ray, T. and Evans, J. (1986) *Post Innovation Performance: technical development and competition*, Macmillan, London.

Giddens, A. (1985) *The Constitution of Society* Polity Press, Oxford.

Gordon, D.M., Edwards, R. and Reich, M. (1982) *Segmented Work, Divided Workers*, Cambridge University Press, Cambridge.

Gouldner, A.W. (1954) *Wildcat Strike*, Free Press, New York.

Grant, R.M. (1991) *Contemporary Strategy Analysis*, Basil Blackwell, Oxford.

Helper, S. (1990) Comparative Supplier Relations in the US and Japanese Auto Industries — an Exit/Voice Approach, *Business and Economic History*, November.

Herrigel, G.B. (1990) "The Politics of Large Firm Relations with Industrial Districts", presented at the Networks Workshop on the Socio-Economics of Inter-Firm Co-operation, Social Science Centre, Berlin, 11–13 June.

Hickson, D.J., Pugh, D.S. and Phesey, D.C. (1969) Operations Technology and Organizational Structure: an empirical reappraisal, *Administrative Science Quarterly*, **14**, 378–97.

Keesing, R.M. (1974) Theories of culture, *Annual Review of Anthropology*, **3**.

Jamous, H. and Pelloille, B. (1970) "Professions or Self-perpetuating Systems" in *Professions and Professionalisation* (Ed. J.A. Jackson) Cambridge University Press, Cambridge.

Johnson, G. (1987) *Strategic Change and the Management Process*, Basil Blackwell, Oxford.

Kanter, R.M. (1983) *The Change Masters*, Simon and Schuster, New York.

Karpik, L. (1972) Les politiques et les logiques d'action de la grande entreprise industrielle, *Sociologie du Travail*, **1**, 82–105.

Kimberley, J.R. (1980) "The life cycle analogy and the study of organizations", *The organisational life cycle*, (Eds J.R. Kimberley and R.H. Miles), Jossey Bass, San Francisco, 1–14.

Kondratiev, N.D. (1927: 1935 edn.) The Long Waves in Economic Life, *Review of Economic Statistics*, 17 November, 105–15.

Kuhn, T.S. (1975, 1962, 1970) *The Structure of Scientific Revolution*, Chicago University Press, Chicago.

Langrish, J., Gibbons, M., Evans, W. and Jevons, F. (1972) *Wealth from Knowledge*, Macmillan, London.

Lash, S. (1990) *Sociology of Post-Modernism*, Routledge, London.

Lee, D. (1982) "Beyond de-skilling: skill, craft and class", in *The Degradation of Work?* (Ed. S. Wood), Hutchinson, London, 146–62.

Lee, G.L. (1986) "The Adoption of computer based systems in Engineering: managerial strategies and the role of professional engineers", presented at UMIST/Aston Organization and Control of the Labour Process: Fourth Annual Conference, April.

Lodge, G.C. and Vogel, E.F. (1987) *Ideology and National Competitiveness — an analysis of nine countries*, Harvard Business School Press, Boston, MA.

Loveridge, R. (1982) "Business Strategy and Community culture", in *The International Yearbook of Organizational Studies*, (Eds P. Dunkerley and G. Salaman), Routledge and Kegan Paul, London.

Loveridge, R. (1983a) "Labour Market Segementation and the Firm", in

Manpower Planning (Eds J. Edwards, C. Leek, R. Loveridge, R. Lumley, J. Morgan and M. Silver), John Wiley, Chichester.

Loveridge, R. (1983b) Sources of Diversity in Internal Labour Markets, *Sociology* **17**, 1 February, 44–62.

Loveridge, R. (1989) "Strategic Frames and Formulae in Decision-making Events", British Academy of Management Workshop on Organizational and Strategic Decision Making, University of Bradford Management Centre, January.

Loveridge, R. (1990) "Footfalls of the Future", in *The Strategic Management of Technological Innovation* (Eds R. Loveridge and M. Pitt), Wiley & Sons, Chichester.

McCormick, K. (1988) Engineering Education in Britain and Japan, *Sociology* **22**, 4 November, 583–605.

Mannheim, K. (1944) *Diagnosis of our Time*, Oxford University Press, Oxford.

Marshall, A. (1920: 1961 edn.) *Principles of Economics*, Macmillan, London.

Merton, R.K. and Kitt, A.S. (1950) "Contributions to the Theory of Reference Group Behaviour", in *Continuities in Social Research*, (Eds R.K. Merton and P.F. Lazarfield), Free Press, Glencoe, Ill.

Miles, R. and Snow, C. (1978) *Organizational Strategy, Structure and Process*, McGraw-Hill, New York.

Miles, R. and Snow, C. (1986) Organizations: new concepts for new forms, *California Management Review* **28**(3), 62–73.

Miller, D. and Friesen, P.H. (1984) *Organizations: a quantum view*, Prentice Hall, New Jersey.

Mok, A.L. (1975) "Is er een Dubbele Arbeidsmarkt in Nederland?" in *Werkloosheid* (Ed. A.L. Mok), Martinus Nijhoff, The Hague.

Nelson, R. and Winter, S. (1982) *An Evolutionary Theory of Economic Change*, Harvard University Press, Cambridge, MA.

Nockolds, H. (1976) *Lucas: the first hundred years* (2 vols), David and Charles, Newton Abbot.

Offe, C. (1985) "Ungovernability: the renaissance of Conservative Theories of Crises", in *Contradictions of the Welfare State* (Ed. J. Keane) MIT Press, Cambridge, MA.

Ogburn, W.F. (1922) *Social Change*, Macmillan, New York.

Ouchi, W.G. (1980) Market, bureaucracies and clans. *Administrative Science Quarterly* **25**, 129–140.

Patel, P. and Pavitt, K. (1988) "Technological Activities in the F.R. Germany and the UK: Differences and determinants", SPRU Designated Research Centre Discussion Paper No. 58, March.

Pavitt, K. (1986) "Technology, Innovation and Strategic Management", in *Strategic Management Research*, (Eds J. McGee and H. Thomas), John Wiley, Chichester.

Pavitt, K. (1988) Commentary on Tushman M. and Anderson P. "Technological Discontinuities" in *The Management of Strategic Change* (Ed. A. Pettigrew), Basil Blackwell, Oxford.

Pavitt, K. (1991) Key Characteristics of the Large Innovating firm, *British Journal of Management*, **2**, 41–50.

Perez, C. (1983) Structural change and assimilation of New Technologies in the Economic and Social Systems, *Futures*, October, 357–75.

Piore, M. and Sabel, C. (1984) *The Second Industrial Divide*, Basic Books, New York.

Polanyi, M. (1964) *Personal Knowledge*, Harper, New York.

Porter, M. (1990) *The Competitive Advantage of Nations*, Macmillan, London.

Rogers,E.M. (1962: 1983 edn.) *Diffusion of Innovations*, Free Press, New York.

Rothwell, R. (1977) The characteristics of successful innovations and technically progressive firms, *R & D Management*, **7**, 3 June.

Rumelt, R.P. (1987) "Theory, Strategy and Entrepreneurship" in *The Competitive Challenge* (Ed. D.J. Teece), Ballinger, Cambridge, MA.

Schon, D.A. (1963) Champions for radical new innovations, *Harvard Business Review*, March/April, 133–60.

Sciberras, E. (1977) *Multinational Electronic Companies and National Economic Policies*, JAI Press, Greenwich, CT.

Schumpeter, S.J. (1939) *Business Cycles*, McGraw-Hill, New York.

Shearman, C. and Burrell, G. (1988) The Structure of Industrial Development, *Journal of Management Studies*, **24**, 4 July, 325–45.

Simon, H.A. (1957) Rational Choice and the Structure of the Environment, *Review of Economic Studies* 20(1), 40–55.

Slade, J.B. and Fordham, P.A. (1990) *European Automotive Component Industry in the 1990s*, West Midlands Engineering Employers Association, Birmingham.

Smith, C., Child, J. and Rowlinson, M. (1990) *Innovation in Work Organisation — the Cadbury Experience*, Cambridge University Press, Cambridge.

Sorge, A. and Warner, M. (1986) *Comparative Factory Organisation*, Gower, Aldershot.

Spender, J-C. (1980) "Strategy Making in Business", unpublished Doctoral Thesis, School of Business, Manchester University.

Teece, D.J. (1977) Technological Transfer by Multi National Firms, *Economic Journal*, June, 242–61.

Teece, D.J. (1987) "Profiting from Technological Innovation", in *The Competitive Challenge*, (Ed. D.J. Teece), Ballinger, Cambridge, MA.

Turnbull, P. (1986) The "Japanisation" of Production and Industrial Relations at Lucas Electrical, *Industrial Relations Journal*, **17**(3), 193–206.

Tushman, M.L. and Anderson, P. (1986) Technological Discontinuities and Organizational Environments, *Administrative Sciences Quarterly*, **31**, 439–65.

Weick, K. (1987) "Perspectives in Organizations" in *Handbook of Organizational Behaviour* (Ed. J.W. Lorsch), Prentice Hall, Englewood Cliffs, NJ.

Van de Vliet, A. (1986) Where Lucas sees the Light, *Management Today*, June, 123–40.

Whipp, R. and Clark, P. (1986) *Innovation and the Auto Industry*, Francis Pinter, London.

Wilkinson, B. (1983) *The Shopfloor Politics of New Technology*, Heinemann, London.

Williamson, O.E. (1982) "Efficient Labor Organization", Conference on Economics and Work Organization, University of York, March.

Williamson, O.E. (1985) *The Economic Institutions of Capitalism: Firms, Markets, Relational Contracting*, Free Press, New York.

Woodward, J. (1965) *Industrial Organizations*, Oxford University Press, Oxford.

6

STRATEGY AND TECHNOLOGICAL LEARNING: AN INTERDISCIPLINARY MICROSTUDY[1]

Mark Dodgson

INTRODUCTION

Rapid and turbulent technological change gives considerable advantages to those firms most capable of dealing with novelty. This chapter discusses the benefits accruing to firms which, in response to the considerable competitive and technological uncertainties engendered by a novel and disruptive technology, place particular emphasis on strategies for learning. It is argued that the differential ability of firms to learn about biotechnology — a "discontinuous" technological change emanating from outside the corporate sector — has created opportunities for new firms to compete with, or at least provide supplementary competences to, large established firms. Using an example of one of the world's leading new biotechnology companies, it is shown how learning has been a fundamental element of its strategy, and how it provides the basis of its comparative competitive advantage. It is argued that in the "fast learning" firm, strategies for technological and human resource development are tightly integrated. The efficiency with which the company searches for, accumulates, diffuses and reviews technological know-how is dependent upon the employment system, organizational structures and managerial abilities utilized around its commitment to technological activities.

CONCEPTUALIZING TECHNOLOGY STRATEGY

There exists a large and growing literature concerned with the relationship between technology and strategies in firms (Porter, 1985; Horwitch, 1986; Teece, 1987; Dodgson, 1989; Loveridge and Pitt, 1990). The breadth of the literature is matched by the range of definitions of what technology strategy is (see Adler, 1988). In this paper, strategy is

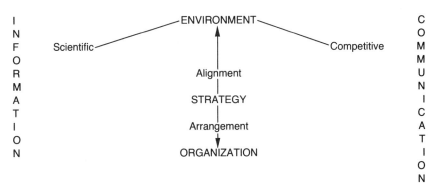

Figure 6.1

conceptualized as an interjacent filter linking a firm's activities and organizational structure with its competitive, technological and legal/ regulatory environment. This model (which is based on Chandler (1962) and Snow and Miles (1983)) is shown in Fig. 6.1. Strategy is a means by which a firm's internal strengths and weaknesses are linked with the opportunities and threats provided by its environment.

In a normative sense, the model describes how, in order to be effective, strategy needs to be aligned to its environment and arranged in line with its competences. The relationship is multi-directional. Although strategy reflects the environment in which the firm operates, it may also exist to change the existing technological and competitive circumstances. Similarly, although the strategy influences the activities and organization of the firm (following Chandler's dictum, structure follows strategy), the capacities of a firm (particularly its knowledge and skills base) affects the strategies likely to be adopted. In order for strategy to be fully reflexive and responsive, effective information and communications channels must exist to inform all levels of the strategy formulation process.

In what follows, this conceptualization of strategy will be examined using a case study. The changing competitive and technological environment facing companies using biotechnology will be described. Then, using a study of a leading biotechnology company, a strategy based on technological learning will be described and analysed, and the factors which have enabled the firm to develop such a strategy will be delineated.

THE TECHNOLOGICAL DEVELOPMENT OF BIOTECHNOLOGY

Three facts about biotechnology are important to the argument developed here. First, it emerged from the science base, from within universities and teaching hospitals, rather than within industry. Second, it is a recent innovation: the two scientific discoveries underpinning its subsequent development, recombinant DNA (rDNA) and monoclonal antibodies (MAbs), were only discovered in the early 1970s. Third, it is a process innovation; it provides the basis for the discovery of new products, but it is not a product in itself.

Biotechnology may be considered a "discontinuous" innovation. In the sense used here, this refers to a discovery having the potential seriously to affect firms' previous paths of technological development. Nelson and Winter (1982) and Dosi (1982) offer the idea of "technological trajectories". In firms, this refers to the way that technology develops incrementally, reflecting endogenous factors, such as the firm's cumulated learning abilities, and exogenous factors, such as market pressures. A discontinuity occurs when the scientific discovery, in this case genetic engineering, profoundly affects existing technological trajectories.

It is not our intention to describe the details of biotechnology here.[2] For the purposes of this paper it is enough to describe biotechnology as the engineering of changes in the genetic structure of micro-organisms, and to emphasize its novelty. It is an interdisciplinary technology, involving a wide range of skills and knowledge; for example: molecular biology, microbial and cellular physiology, and biochemical engineering. It is a technology which is qualitatively different from what went before. As such it cannot easily augment previous know-how and practices. Hence the "discontinuity".

In the literature on discontinuous innovation (see particularly, Sahal (1982)), these radical changes are assumed to create technologies with marked performance and price advantages over existing technologies. A further element of the literature distinguishes between discontinuities which are "competence enhancing" and "competence destroying". Competence enhancing discontinuities are "order-of-magnitude improvements in price/performance which build upon existing know-how". Competence destroying discontinuities are "so fundamentally different from previously dominant technologies that the skills and knowledge base required to operate the core technology shift" (Tushman and Anderson, 1987). Biotechnology would appear, therefore, to have the characteristics of competence destruction (Hamilton *et al.*, 1990). It emerged from outside the corporate sector, and hence was not based on its accumulated scientific competences. It involves novel skills, and it has the potential to provide price/performance advantages over previous technologies. However, the temporal element in the innovation is all-important, and has to be taken into account.

According to Pavitt (1987) "The commonly held assumption ... that new vintages of technology immediately reach economically superior performance is rarely borne out in practice. A period of trial, error, learning and associated incremental change is necessary before a major new technology begins to reach its potential." Although the new skills of biotechnology are not competence enhancing, they may only prove competence destroying once the technology is proven. The different vintages of technology may exist alongside one another. Indeed, as a new technology is gradually developed, there may well be elements of convergence between it and the older technology.

Biotechnology has certainly not realized the over-inflated early expectations of it. In part, this reflects the time-scale involved. Semiconductors took at least twenty-five years to develop into the range of applications we know today. The two major biotechnology break-throughs of rDNA and MAbs — process inventions with no immediate products — only occurred in the mid-1970s. The technology is still very young, and its long-term impact on the skills and knowledge bases of firms remains to some extent uncertain. What is clear, however, is that the technology has caused some considerable adjustment in the R and D activities of firms. Two aspects of this adjustment will now be considered: the formation of a new type of firm, the dedicated biotechnology firm (DBF), and the high levels of collaborative activity between established firms and DBFs.

The development of dedicated biotechnology firms

DBFs began to emerge in the United States in the early to mid-1970s. It is estimated that there are over 400 such firms in the USA, and over eighty in Europe (OTA, 1988; Orsenigo, 1989). The largest DBF, Genentech in the USA, 60 per cent of which was recently acquired by a large European drugs company, has sales in excess of $300 million, employs over 1750 people, and is valued at $3.8 billion. Figure 6.2 shows the growth in the number of DBFs in the USA and Britain. It shows how, since the early 1980s, the number of DBFs being formed has declined. Some leading DBFs, however, have enjoyed rapid patterns of growth.

The emergence of DBFs in the USA has been attributed to a number of interrelated factors. Important among these have been: the development

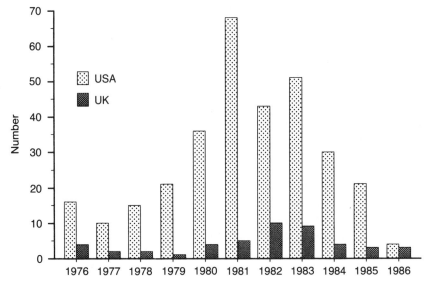

Figure 6.2: Number of new biotechnology firms established in the USA and UK
(1976–86); (From OTA 1988; Oakley et al., 1990.)

of novel scientific techniques in universities and research hospitals; the
ready availability of venture capital; the entrepreneurial drive of
scientists, venture capitalists and intermediaries between the two; and
the synergistic relationships DBFs have formed with large firms.

Hall's (1987) account of the formation of Genentech, a company
initially created by an alliance between a venture capitalist and some
entrepreneurial scientists, highlights some of the factors which have led
scientists to work for or with DBFs. Naturally the possibility of large
financial rewards through the high salaries and stock options which a
fast growth company can provide have been important. Equally, the
opportunity to undertake well-funded research, not dissimilar to the
research undertaken in academia yet avoiding the vagaries of decision-
making by public grant-giving bodies, has been an attraction. The
informal environment of a young, small firm, more familiar to academics
than the bureaucratic arrangements often found in large established
firms, has also been attractive.

The relationships between DBFs and large, multinational companies
are one of the major features of the development of biotechnology.
Large, established firms were generally slow to become involved in
biotechnology, but currently are devoting considerable resources to it,
and many have acquired DBFs. Some large firms were involved in

biotechnology very early on — for example, ICI in Britain, and Eli Lilly in the USA. However, in the early and mid-1980s the majority of industrial activity in biotechnology took place in DBFs, with large, established firms tending not to invest heavily, but keeping a "watching brief" on DBFs' efforts. Indeed, the 1988 OTA report describes a survey which concluded that, as late as 1987, DBFs played the major role in industrial biotechnology R and D, spending $1.2 billion on R and D compared to the large corporations' $0.8 billion. The report suggested, however, that major corporations' expenditure was increasing comparatively more rapidly.

There are a number of other reasons why DBFs, rather than large firms, initially attempted to develop biotechnology. One reason lay with the unproven and risky nature of the technology itself. If DBFs were developing the technology, and undertaking the risk, why should large firms bother developing competences which they could easily acquire when the science was better established, and market opportunities were clearer? Also, the conventional pharmaceutical companies' R and D structures were primarily arranged in line with the disciplines of biology and chemistry. The new skills of molecular biology did not accord well with these existing structures. Furthermore, the traditional methods used in the search for new drugs — the search for new substances, then extensive screening for potential applications — were very different from the possibilities provided by biotechnology, whereby drugs could be "designed" with a particular application in mind. As Dosi (1989) puts it, "What the firm can do technologically in the future is narrowly constrained by what it has been capable of doing in the past." For some large firms, biotechnology was perceived, therefore, as a competence destroying discontinuity.

Another reason lies within the focus of this chapter, namely the differing nature of the organization and management of DBFs compared with large firms. Radical new technologies, like biotechnology, with the potential to alter existing structures and systems profoundly, put a premium on firms with the ability to learn rapidly about new scientific and market possibilities, and the personnel and organizational structures to take quick advantage of these lessons for commercial benefit. It was the DBF which, being new, had the opportunity to create novel management systems and practices and organizational arrangements in order to take advantage of the opportunities biotechnology provided.

The relationships between the science base, DBFs and large firms

One analysis of the relationship between academia, DBFs and established firms is based on the work of David Teece. Teece suggests that for firms to realize the full return from their investment in creating a technological innovation, they need to access what he calls "complementary assets" (Teece, 1987). In biotechnology such assets might include competitive manufacturing, marketing and distribution networks, and the ability to deal with the regulatory procedures involved in getting new products on to the market. It is argued that the science base, DBFs and large firms have in the past possessed distinctive advantages in the ability to create biotechnology innovations and access complementary assets, although this is changing (Pisano et al., 1988). Initially, scientific expertise lay in the science base, and was gradually transferred to DBFs as they developed their own R and D competences. Large firms initially lacked the R and D competences to innovate in biotechnology themselves, but they had all the complementary assets necessary to commercialize discoveries, such as manufacturing skills, marketing and distribution networks, and experience with regulatory processes. These distinctive organizational advantages provided the basis for considerable amounts of collaboration between the three groups. Over time, however, larger firms began to develop their own R and D skills in biotechnology, and some of the major DBFs began to manufacture and market their own products, so the picture becomes more complicated.

A number of features of collaboration in biotechnology have emerged:

1. Collaboration is common between the new biotechnology firms (DBFs) and established firms. Pisano et al. (1988), taking a random sample of 200 biotechnology collaborative agreements, found 62 per cent were between DBFs and established firms; 10 per cent were between established firms; 5 per cent were between DBFs; and the remainder were with universities and research laboratories. This supports the contention of the differing organizational capabilities of the actors involved in biotechnology. A study reported by Yarrow (1989) showed that in 1987 the ten largest US pharmaceutical companies had an average of three partnerships with DBFs.

2. The most common form of collaboration is in R and D. Pisano et al. found the most common function involved in collaboration to be R and D (36 per cent of cases). Hagedoorn and Schankenraad (1990), in a sample of 1213 biotechnology collaborative agreements reported in the press, also found joint R and D to be the most common mode (29.8 per cent).

3. The motives for collaboration vary. In a sample of 638 joint biotechnology R and D collaborative arrangements, Hagedoorn and Schankenraad (1990) found the three most common motives to be technological complementarity (38.1 per cent), reduction of innovation lead times (31 per cent), and lack of financial resources (12.1 per cent).

They provide, therefore, a mixture of strategic and tactical concerns. Technological complementarity may be a long-term strategic motive, whereas reducing lead times for particular innovations, or raising finance (by DBFs), may be more immediate concerns. The strategic element of these collaborations may be a reason why, according to Orsenigo (1989), there is a considerable stability in many large/small firm links.

Apart from the increasing expenditure by large firms on biotechnology, many of them have acquired DBFs. In addition to Hoffman-La Roche's majority purchase of Genentech, Eli Lilley has acquired Hybritech, Bristol-Myers has acquired Genetic Systems, and Glaxo has purchased Biogen's European laboratory.

Celltech Ltd: an exemplar DBF[3]

Formed in 1979, Celltech is Europe's leading DBF, and dominates among British DBFs. In 1989 it had sales of just under £20 million, and employed nearly 500. It is the world leader in the bulk manufacture of biotechnology products such as MAbs, and presently has a number of drugs in development, including two which the company is targeting in an attempt to market them within a few years. The company has followed the development pattern of a number of leading US DBFs in first creating a research and development business, then a manufacturing capacity, and finally building on these by attempting to become an integrated biopharmaceuticals business, developing, manufacturing and marketing drugs (see Hamilton *et al.*, 1990).

Its comparative advantage over large pharmaceutical firms originally lay in its linkages with the science base, in particular with the UK's Medical Research Council. As it began to develop its internal skills, it identified the scientific feasibility and commercial potential of the manufacture of biotechnology products. It rapidly developed a world-leading capacity in its manufacturing facilities, ahead of its major competitors. More recently it has been attempting to integrate its activities fully on the basis of its R and D base, its manufacturing know-how, and the belief that it can identify market niches which it can fill quicker and more cheaply than its large firm competitors. This latter strategic move is, of course, a risky endeavour.

Not many smaller firms can attempt to compete with the world's most powerful multinationals. The decision to embark on the strategy followed very extensive discussions, and was made in part because it believed that it had certain comparative competitive advantages over large firms. These included the way it could operate differently from established firms; to quote from its Board minutes: ". . . avoiding some of their overheads, their bureaucratic structures and their internal politics". Furthermore, it argued that large firms had continued to use traditional methods of drug discovery, without realizing the full potential of the advances in molecular and cell biology.

Also important in this respect was the way large firms were acquiring DBFs, which was believed to indicate that they found it easier to acquire external competences than to adapt their existing organizations. Celltech's strategy therefore reflects its differential ability over large firms to respond quickly to technological and competitive circumstances. The strategy is informed by, and focuses on, its learning processes. As the strategy alters, so too does the focus of its learning processes. Thus the original strategy required priority to be placed on building R and D skills. Subsequently emphasis was placed on manufacturing competences. Currently the company is adding to these skills a growing ability in its medical, regulatory affairs and marketing functions.

In this respect Celltech has pursued an adaptive business strategy, amending its aims in line with changing technological and competitive circumstances on the one hand, and its own developing competences on the other. The key to its rapid development of scientific and managerial competences — and hence its ability to comprehend and benefit from the rapid changes in its environment — lies in its strategy of high dedication of resources to R and D, and its sophisticated human resource strategy.

This chapter will not describe Celltech's R and D and human resource strategy in detail, as this is done elsewhere (Dodgson, 1990a). Instead, the aim is to analyse the role of learning in the company's strategy, by referring to some elements of these strategies.

A BRIEF REVIEW OF THE LITERATURE ON LEARNING

Technological and organizational learning in firms is addressed by a wide range of academic literatures, such as innovation studies, industrial economics, organizational behaviour, and management studies. A number of examples illustrate the ubiquity of concepts of learning.

Learning is central to Cyert and March's theory of the firm. "A theory of long-term behaviour in organizations must contain a theory of how organizations learn, unlearn, and re-learn" (Cyert and March, 1963). The differing ability to stimulate and accumulate technological knowledge in changing technological and competitive circumstances is a feature distinguishing Burns and Stalker's classic (1961) analysis of organic and mechanistic organizational forms. Organizational learning is the focus of Argyris and Shon (1978), and is a feature of Ansoff's (1982) contribution from a strategic management perspective.

Concepts such as the learning curve are widely used in managerial literature, and its use as an analytical tool is extensive, actively promoted by consultancy companies such as the Boston Consulting Group. Learning features in some managerial literature on new product development (Maidique and Zirger, 1985), and recently economic theory's consideration of technological and industrial development has begun to emphasize the importance of learning: "... the features of the evolution of each industry are ... 'ordered' by the patterns of learning, and by the ways the latter influence the competitive process" (Dosi, 1989). Arthur's new positive feedback economics also emphasizes the importance of learning in knowledge-based industries (Arthur, 1989).

Numerous stimuli to learning in firms have been suggested, including both environmental and intrinsic factors. Freeman and Perez (1988) ascribe renewed learning activity in firms — about organizational forms and the skills required — to the environmental turbulence engendered by changes in "techno-economic paradigm". Triggers to learning within firms include existing problems (Hedberg, 1981); affluence (Cyert and March, 1963); and executive succession (Tushman et al., 1986).

The literature on technological learning has been marked by considerable heterogeneity, often compounded by definitional differences. It has tended to focus on learning in production and post-production, rather than on R and D, and has concentrated rather narrowly on the learning that occurs as a "natural" result of cumulated experience. To some extent this emphasis on manufacturing learning follows the impact of the work of Arrow (1962), and his contention that firms develop increasing skills in production, with the result of reducing real labour costs per unit of production ("learning-by doing"). While this thesis has its shortcomings, it has attracted considerable attention, particularly amongst economists. Another feature of the literature on learning is that it focuses primarily on established rather than newer firms.

Perhaps the most sophisticated analysis of technological learning from an economic/innovation studies perspective is provided by Rosenberg (1982). He focuses his analysis mainly on learning-by-doing and learning-by-using (learning from the use of a product), but he also addresses learning in R and D:

> R and D is a learning process in the generation of new technologies. . . . At the basic end of the spectrum, the learning process involves the acquisition of knowledge concerning the laws of nature. Some of this knowledge turns out to have useful applications to productive activity. At the development end of R and D is a learning process that consists of searching out and discovering the optimal design characteristics of a product. At this stage, the learning is oriented toward the commercial dimensions of the innovation process: discovering the nature and combination of product characteristics desired in the market . . . and incorporating these in a final product in ways that take account of scientific and engineering knowledge.
>
> (Rosenberg, 1982)

Thus learning is not uniform in purpose or nature, even within functions. Learning has a strategic element, characterized by search routines designed to bring knowledge into the firm, or to generate it internally, about a scientific field or a technology. It also has a tactical element, which essentially is concerned with the provision of a particular product or service to a customer. Learning is at once concerned with science and commerce.

With reference to the literature cited above, and for the purposes of this paper, learning is defined as the ways firms build and supplement their knowledge bases about technologies, products and processes, and develop and improve the use of the broad skills of their workforces.

Conceptualizing learning

Learning is a heterogenous concept, it has differing aims and foci and involves different processes.

The aims of learning

As Harris and Mowery (1990) tell us:

> When firms generate innovation, or when they adopt innovations generated by others (and whether they act deliberately or "accidentally") there is, at minimum, the opportunity for learning. In other words there are two separable outcomes of innovation processes, the innovation itself and the learning about the innovation and innovation process.

In this way the aims of learning may be strategic or tactical.

Tactical learning may be described as that which has an immediate problem-solving nature. It may relate, for example, to a product or to an operational problem. The learning required may involve recourse to an R and D or engineering department, or to an external source, perhaps through the use of a consultant, research association, university or even collaboration between firms. The aim of the learning is identifiable, and the time-scale is short and prescribed. Any improvement in the innovative capabilities in the firm are project specific. Strategic learning extends beyond immediate issues and involves firms developing managerial and scientific/technological skills and competences which provide the basis for future, perhaps unforeseen, projects. Strategic learning attempts to maximize the returns from the opportunities that the innovation process provides.

In the sense used here, strategic and tactical learning are not synonymous with research as opposed to development, as both these activities may be strategic or tactical in focus. Strategic and tactical learning are more closely analogous with what Patel and Pavitt (1988) refer to as dynamic and myopic firms. The latter category is concerned only with the immediate problems, the former takes the broader, long-term view of technological accumulation: that is, technology is acquired for its potentiality. Argyris and Shon's (1978) analytical distinction between single-loop and double-loop learning and deutero-learning is also useful. In a very simplified way, single-loop learning involves feedback into current decision-making practices, so as to improve future decisions. Double-loop learning involves organizations questioning the whole basis of their decision-making practices and involves wholesale modification within the company. Deutero-learning examines the whole process of learning within the company, and can be viewed as "learning about learning". Tactical learning is equivalent to single-loop learning. Strategic learning includes both double-loop and deutero-learning.

The foci of learning
Learning does, of course, have different foci. These may include R and D, production, and managerial/organizational concerns, but this list is nowhere near exclusive. Learning is important throughout the activities of a firm. The foci of learning change over time. As Hamilton, *et al.*'s (1990) study of a sample of eighty-seven US DBFs shows, these firms' activities steadily shifted from science to commerce over time. The foci of learning adapt accordingly.

Two particularly important areas of learning are understanding environmental change and developing managerial competences. A central focus of strategic learning is gaining an understanding of the opportunities provided by changes in the firm's competitive and technological environment. In the case of DBFs, and in all leading technology companies, this requires the detection of technological shifts which threaten existing ways of doing things and provide the potential for alternatives. One of the major obstacles to successful growth in high technology firms, as described by Hambrick and Crozier (1985), is the recruitment of people who, with continuing training, will grow into their new and increasing responsibilities as the company grows. Management learning is a crucial activity.

Learning process
Technological learning differs broadly in its nature. Three broad kinds of learning are distinguished here: search, accumulation and diffusion, and review.

Search involves seeking knowledge, either internally through the firm's own activities, for example in R and D or engineering, or externally, for example from scientific and technological scanning. Internal search may be directed into discrete technologies or projects, and may also include efforts to find linkages between different scientific and technological activities. Search activities may be strategic or tactical; they may be concerned with a particular problem to be solved or may be directed towards the long-term acquisition of potentially useful knowledge.

Once information has been sought and found, it then needs to be disseminated into the firm's stock of knowledge. Accumulation of know-how again has strategic and tactical elements. Learning about a tactical problem should be accumulated and diffused within the firm, so as to prevent repetition of search activities. Learning about strategic issues should be accumulated and diffused to inform future strategic plans and departures in the various parts of the firm.

An important aspect of learning is reviewing what you know. Such reviews can assist strategic decision-making, and can highlight deficiencies in the search and accumulation and diffusion process. An aspect of review learning is learning from mistakes or failures.

LEARNING IN CELLTECH

Before starting to describe how Celltech learns, it is perhaps appropriate to provide some evidence on how successful it has been at learning. A number of indicators will be suggested, and these should be considered in the context of the company's rapid pattern of growth. It should be noted that any indicators of this nature are only partially meaningful, and can only be considered as proxy measures.

Growth in patent activities
The number of patent applications made by Celltech is shown in Fig. 6.3. The company registered its first patent, based on its own research, six months after its formation. It currently has more biotechnology patent applications than do Wellcome and Glaxo, and almost as many as Beecham. In the first five years of Celltech's existence, slightly more of the twenty-five patent applications came from external sources, such as the Medical Research Council and universities, rather than being based on internal R and D. Since 1985, many more applications based on internal R and D have been made (fifty-one internal to twenty-nine external).

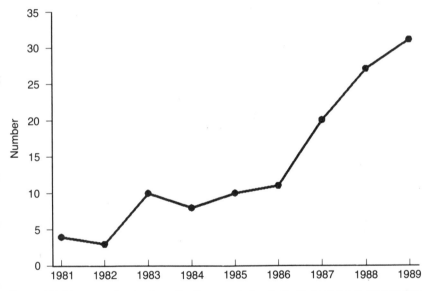

Figure 6.3: Number of patent applications made by Celltech (individual discoveries; many are multi-registered).

Growth in manufacturing capacity
This is suggested as an indicator of learning abilities, as the company is at the world forefront of the development of this technology. Bulk production (i.e. production of batches of several grammes) is an extraordinarily complex process, with contamination a permanent threat. Celltech has pioneered much of the development of this technology, and therefore its expanding capacity may be used as a proxy

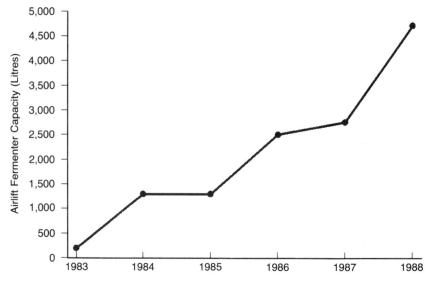

Figure 6.4: Cumulative growth in manufacturing capacity at Celltech.

for technological learning. Figure 6.4 shows diagrammatically how rapidly this capacity has grown.

Learning in R and D

The basis of Celltech's technological learning lies in its high commitment to R and D. In 1989 the company spent over £10 million on R and D, equivalent to 51 per cent of turnover. £6.2 million of this was own-account expenditure (rather than sponsored R and D), accounting for 31 per cent of sales turnover.

The company's initial R and D strategy was very broad-based, and unlike that of many of its competitors, involved activities in both rDNA and MAbs. In the event, this proved extremely fortuitous, as the two technologies unexpectedly converged to some extent in the guise of antibody engineering. The consolidation of Celltech's expertise in the two technologies of large-scale manufacture of proteins in cell culture and antibody engineering has provided the company with its core technological competences and the basis of its competitive position. It enabled the company to embark on its antibody-based drug discovery programme.

The company's selection of technologies in R and D has proved highly effective. The technologies in which the company was active in 1983 remain important in 1990. Astute selection of technologies, in circumstances of considerable uncertainty, has focused the company's search activities and enabled it to develop depth in its knowledge base. In the past, the company was less successful in its project selection; a number of projects failed to realize their immediate objectives. In part this reflected the company's early learning focus on scientific rather than commercial issues. Following these failures, reviews were undertaken of the projects, and the lessons learnt affected subsequent project selection and management.

Learning from internal R and D is complemented by extensive learning from R and D linkages with external bodies, both private and public sector. Collaboration, as we have seen, has been a feature of the development of biotechnology. The company undertakes considerable amounts of contract R and D, and has collaborated with many of the leading US, Japanese and Continental European pharmaceutical firms. In its collaborations, the company has attempted to maximize the amount of "enabling" technology which is required in order to undertake the contract. Such technology remains proprietorial to the company after the collaboration. An example of this strategy is provided by a major contract for the US drugs company, American Cyanamid. In this programme, designed to develop aids to cancer diagnosis and treatment, Celltech has retained all rights to non-cancer discoveries. It has formalized its expectations regarding the development of enabling technology. In the past, however, building up enabling technology has often been a problem. A balance has had to be retained between meeting short-term (tactical) aims, in order to earn income, and long-term (strategic) objectives of building the company's science base. In practice, managers have had to be extremely flexible in conflating these long and short-term demands. Potentially this tension between long- and short-term needs could manifest itself in disputes between line managers and project managers. To circumvent these, and to ensure that enabling technology is given a high priority, project managers are not asked to deal with the allocation of resources between project and enabling technology. Such decisions are referred upwards in the organization; the bigger the decision, the higher it is referred, and the more likely that the outcome will reflect strategic rather than tactical concerns. Decisions like these, on the selection of projects, are an important focus of learning.

The company also has very extensive collaborations with academia. Since its formation, Celltech has always had a senior manager responsible for academic liaison. At present, this activity is undertaken by the Director of Academic Liaison who is a member of the company's

Strategic Management Group. Celltech has also always had a Science Council, an academic consultative group consisting of leading UK scientists. Included among the functions of the Science Council are the bringing of new ideas to Celltech, and help in filtering new ideas coming from any source; and responding to requests from R and D management for reviews of major topics (e.g. large new project proposals). It provides a useful supplement to the company's search and review activities.

In 1988 £1.25 million was spent by Celltech on its academic collaborations. Most of these collaborations tend to be longer term than simple licensing agreements. Typically, they involve Celltech funding a laboratory or individual researchers (usually for three-year projects), in return for the assignment of exclusive rights of any discovery to Celltech. In 1988 around twenty three-year collaborative and consultancy agreements were negotiated by Celltech.

Appendix 1 gives some indication of the extent and range of collaborations Celltech had undertaken by 1989. Over one-third of Celltech's collaborations have been with overseas partners. Celltech is involved in a variety of forms of academic liaison, including a number of government and research council research initiatives. Appendix 2 includes some details of the nature of some of Celltech's current collaborations. The value to Celltech of academic collaborations is seen in the extensive involvement of academia in its major current and past product and process development activities.

Celltech's links with the Medical Research Council have been particularly important. The MRC provided the company with its first products, and subsequently provided a substantial proportion of its science base. Although not as important as it once was, collaboration with the MRC remains crucial to the company in certain areas, such as the development of antibody engineering.

Learning from making

As we have seen, Celltech's world-leading manufacturing capacity has developed rapidly, involving the company in the creation of new production techniques. An example from Celltech's manufacturing division highlights the importance of learning for the company. The first is in fermentation, the second in roller bottle processes. The manufacturing division has grown rapidly since 1986: then it employed a staff of thirty-five (including quality assurance personnel), and it presently employs 118 (excluding quality assurance personnel). Celltech's original airlift fermenter used in production was purchased in 1983, and proved to be of poor design and quality: it was supposed to have a 100 litre capacity, but could actually only manage 80 litres. Later that year another 100 litre fermenter was bought, and this produced the whole 100 litres. In 1984, both these fermenters were re-designed and modified, and both managed to produce 110 litres. The first 1000 litre fermenter was commissioned in 1984, some eight months after it had originally been planned. Another 1000 litre fermenter was introduced in 1986. This was fabricated entirely off-site, and was placed in the company's purpose-built manufacturing building. This later 1000 litre fermenter took around four or five months to commission.

There were major technical problems with these fermenters. When a Manufacturing Director was recruited in 1986, he inherited a high failure rate in batches (up to one in three failed), caused by microbial infections. These failures were exacerbated by plant teething problems, and some major failures occurred in the new building, such as the boiler breaking down. To deal with these contamination problems, the Director began to review the whole process systematically. He found that the plant was perhaps only 95 per cent right in its design, and needed modification. A multidisciplinary team was established to look at these failures, and a microbiological survey was undertaken on how to improve the system. Eventually the contamination problems were eradicated, and while occasional production problems have continued, these have had more to do with the complexity of the cells being used, rather than equipment problems.

The way that the lessons from these experiences have been learned, and used effectively, can be seen in the example of the company's latest fermenter. This fermenter, with a 2000 litre capacity, has been designed completely in-house. Planning for this piece of equipment began in spring 1987, and it was commissioned a year later. Since its commissioning this fermenter has worked continually without major problems.

Pisano *et al.* (1988) argue that new biotechnology firms "lack the key input necessary for competitive large-scale bioprocessing, namely, production know-how gained through experience". Celltech provides an exception to this rule, and has done so through its ability to learn quickly. Celltech's development of manufacturing processes continues, and it has enjoyed some recent success in developing a microcarrier-based unit process for growing anchorage dependent cells.

The employment system and learning

Effective technology strategies require a considerable degree of confluence between the technological elements involved and human resource development strategies (Senker and Brady, 1989). This applies to virtually all elements of technology strategy, including R and D collaboration (Pucik, 1988). Celltech's top management places great emphasis on policies designed to recruit and motivate talented company employees. This has been a key feature of strategy right from the company's inception.

Celltech's employees are characterized by a number of exceptional features, including high educational qualifications and young age profiles. At the beginning of 1989, the company employed eighty-eight PhD scientists, and 131 graduates or postgraduates. Over 55 per cent of the company's staff, therefore, were graduates or had attained post-graduate degrees. At the same time, three-quarters of Celltech's employees were under thirty-five years of age, and only 12 per cent of employees were over forty.

The company attempts to provide considerable training support to enhance career development. Celltech is deeply involved in graduate and postgraduate training. Apart from twenty-seven Science and Engineering Research Council CASE studentships in 1989, there were five people studying for PhDs while working for Celltech. Celltech's Personnel Manager described the major problem of recruitment in Celltech as finding people sufficiently broad-based to match their scientific and technological skills with the personality to work, and develop greater future responsibility, within a commercial environment. In a fast growing firm, recruitment obviously has to consider not only immediate competences, but also potential future capabilities. One of the attractions Celltech might hold for recruits is that its fast growth offers promotion opportunities.

Trade unions are not represented in negotiations at Celltech. This gives the company's personnel function considerable leeway in its management–labour relations. There is no system of job grading at Celltech; staff are graded individually. Despite the hugely time-consuming process of individual assessment, this system is preferred as it enables considerably greater flexibility in allocating rewards. Celltech operates three share option schemes. An additional incentive scheme involves quarterly bonus awards for those considered to have made exceptional contributions. There is a formal appraisal system under-taken for all staff to inform an annual review undertaken in July of each year. The appraisal is compiled by each employee's superior, with the active involvement of the employee. Every employee must see, and agree with, his or her appraisal. Quarterly objectives are agreed, and past performance in achieving these objectives assessed. Emphasis is placed on rewarding enthusiasm and commitment, and the company has considerable flexibility in its employment system to allow this.

The company's employment system is similar to that identified by Morgan and Sayer (1989) in their study of electronics firms in the M4 corridor, where Celltech is situated. It includes an emphasis on training, intensive screening of recruits, encouragement of self-motivation, and a flexible payment system based on individual assessment. Traditional industrial relations concepts such as "discipline" and "control" are essentially meaningless, as work pace is determined more by individual motivation and peer pressure than by organizational sanction.

Organizational structures and learning

Celltech is a knowledge-intensive company — labour costs in 1988 accounted for 38 per cent of sales turnover — and the principles underlying its organizational structure reflect this. A feature of the company's thinking on creative organizational structure is the concept of "compartmentation", a term from cellular biology describing partial closure from the environment, which allows multi-order feedback to develop between the components of a system. (For a description of how the principles of compartmentation are applied within Celltech, see Dodgson (1991).)

A major method for ensuring scientific excellence within R and D was introduced in 1988, with the creation of the Principal Scientist grade of employee. This innovation was brought in to allow appropriate financial reward for those people wishing to concentrate on developing their scientific expertise, rather than undertaking management responsibilities. The Principal Scientists are appointed following an assessment which includes the use of external academic referees, and they are expected to have, and to build upon, international scientific reputations.

One of the mechanisms that Celltech has used to stimulate creativity has been the sanctioning of scientists spending up to 10 per cent of their time on their own projects. In its earlier years this was an explicit policy, whereas currently the practice is less formally constructed. It has always been necessary for those wishing to undertake these so-called "preliminary studies" to have the approval of their Head of Department. It is still possible for scientists to undertake these studies, which may on occasion exceed 10 per cent of their time. The intention underlying these arrangements is to allow scientists flexibility to use their knowledge in a direction unconstrained by project requirements. It is considered important for the company to provide enough time in the face of commercial tactical demands to allow scientists to develop their know-how to be useful for a wider number of projects than their immediate one. In practice, this has been difficult to arrange, and there have been tensions between short- and long-term pressures.

Learning and information and communications systems

As previously suggested, information and communication are essential elements of strategy. They are critical to the learning processes of search, accumulation and diffusion and review. Celltech places great emphasis on its information and communications systems, which involve a mixture of the formal and informal. A number of strategic and operational committees direct the company's activities; the level of formality of these committees might appear unusual for a medium-sized company. However, consideration has to be made of the purpose of these structures, which are not only, and in some cases not primarily, decision-making bodies, but important mechanisms for information-sharing and consensus-forming. From its inception, it has been the policy of Celltech's leading managers to enfranchise as many suitably qualified and experienced people into the decision-making process as possible. The company's "paper-intensiveness" — minutes, agendas, discussion papers, and plans tend to adorn managers' and scientists' desks — is also an important element of information sharing. One of the strengths of having good and broad communications paths between the strategic and operational elements of the company (managers often have responsibilities in both areas), is that each is informed by the decisions and actions of the other.

One particular facility designed to improve communications within the company lies in its new R and D building. In this building is an extended meeting area, where staff can congregate at any time of day. The company has an extensive seminar programme, with both internal and external speakers. During the course of a year, around fifty seminars are given by Celltech staff, and another fifty by invited external speakers. There are also a range of intra- and inter-functional discussions and meetings.

In addition to these formal mechanisms, there is a great amount of informal communication within the company. Senior managers spend a significant proportion of their time talking to managers, scientists and other staff. The practice whereby managers devote considerable effort to knowing their staff is seen to be genuinely believed in, and effectively used, by the senior directors.

A key role in the company's learning processes is played by the Information and Library Service (ILS). This service is provided by seven staff, who regularly provide a range of information to company employees. The ILS keeps around 350 journal titles and has access to numerous online scientific and business data bases; it also has a medical data base on CD ROM, as well as a number of data bases related to particular projects. One project, for example, has a data base of over 700 papers, one other project has over 1000 papers stored. The project data bases include information on patents, science and commercial information, and academic liaisons. More of these data bases are to be created in the future. Each month, patent information is requested on average around forty times, and around forty to fifty online searches are conducted. Three or four requests for company profiles to be undertaken are made each month. The ILS keeps annual reports and market surveys on all its main competitors, and keeps full files on around another dozen companies in which it is interested. Information is also kept on the world top forty biotechnology companies.

In addition to requests on patents from scientists, the company also acts proactively in its information search through its patent department. This department of four staff oversees the company's patenting interests, and has one member who has a permanent brief to watch the patenting of competing organizations. To the formal mechanisms of raising technological awareness discussed above must be added the informal ways in which scientists learn, through attending conferences and visiting other laboratories, and from the high number of academic visitors to Celltech.

CONCLUSIONS

This chapter has presented a mixture of analysis and description. By describing the development of biotechnology and its discontinuous nature, and by conceptualizing strategy and learning in firms, a case has been made for placing learning centrally in the analysis of the industrial and technological progress of a discontinuous technology. It has been argued that the differential ability of firms to learn about, and benefit from, radical technological change has been an important factor accounting for the formation and growth of DBFs, and the high levels of collaborative activity between DBFs and large, established firms. The extent of the learning activities of an exemplar DBF has been described, and the degree to which its competitive position depends on its learning abilities has been discussed. It is the intention of this final section of the chapter to scrutinize one particular aspect of learning in circumstances of rapid technological change: namely the management problem.

A number of the most important management issues concerned with technological learning have already been highlighted. These include detection of environment changes in technology and competition, knowledge in selection of appropriate technologies and projects, the ability to learn from R and D and from manufacturing, and the ability to manage the processes of learning on the basis of creative organizational structures and effective human resource management. The management challenge is succinctly described by Fildes (1990), President and Chief Executive Officer of Cetus, a leading US DBF:

> The first challenge is managing technical excellence. A company in this field, and especially a small company, really only has one ace in the hole competitively speaking. That ace is its technical prowess. ... So how do we compete with them [large firms like Merck and Lilly] now that they are in biotechnology too? We do it by attracting, motivating, and retaining the best and the brightest, giving them a piece of the business through stock options and providing them with an environment where they can express their creativity in a way that cannot as yet be found in the big companies. The challenge for the manager is nurturing this potent yet fragile environment during a period of rapid change and growth.

A DBF cannot attempt to match the level of resources that a large, pharmaceutical firm can devote to R and D. If a technological lead is to be established, it has to be done on the basis of its behavioural advantages over large firms. These may focus on well-targetted and efficient search activities, rapid and extensive accumulation and diffusion of know-how to improve productivity and reduce replication of activities, and continual review of successes and failures. The emphasis lies in learning faster than do large firms. This places particular priority on organizational and human resource issues.

The centrality of these "soft" issues is shown in an important aspect of the management of R and D: the matching of internal and external learning. Celltech has depended on both, and although both are important, the balance between them varies over time. A crucial aspect of the successful management of strategy and learning is the way that complementarities between external and internal learning are achieved. Organizational structures must facilitate efficient integration of external know-how, and individual employees must be similarly receptive. Decisions have to be made on the sources of learning, to avoid repetition and to maximize the comparative advantages of research in-house and externally. The latter decision requires consideration not only of where technological advantage lies, but also of how future outcomes of research are to be appropriated, i.e. who owns the intellectual property rights. The management skills involved in collaboration are extensive and crucial.

The emphasis placed by top DBF managers on personnel-related issues is understandable, given the nature of the competitive environment in biotechnology. Distinctive scientific and technological knowledge and skills are the key to competitive advantage. Broadening and deepening the knowledge and skills bases through innovation, so as to enable rapid responses to changing technological and competitive environments, are an essential element of strategy. As we have seen, various methods may be used to build the base, in both internal and collaborative activities. An implicit aspect of the strategy, and of the methods used to realize it, is that learning effects are proportional to the size, efficient management, and innovatory activities of the knowledge base. As Harris and Mowery (1990) put it, "With successive innovations comes an increased ability to innovate and manage innovation. Put differently, 'learning curve effects' are especially important to the management of innovation because it is so knowledge intensive."

The analysis of learning presented here focuses particularly on newer, smaller firms at the technological forefront. How applicable, then, is the analysis to the wider population of industrial firms? It is, of course, unwise to generalize from case studies, and DBFs are unusual organizations. Nevertheless, as exemplars of constructions designed to deal with rapid technological change, they may be indicative of the sort of structures and processes necessary. There are obvious differences between such firms and large, established firms. Past learning activities do not constrain DBFs, that is, there is no need for forgetting previously accumulated knowledge and learning processes. Large firms may possess the advantage of past experience in dealing with technological discontinuities: transformations which newer firms have yet to face. Small size has advantages of easier communication and closer adherence between managers and scientists. However, learning advantages do not always lie with smaller firms. Large firms have the greater resources, and may also be less reliant on individuals. Small firms, such as DBFs, may depend particularly on a number of highly knowledgeable people, whose departure would profoundly affect the "corporate memory" of the results of previous learning activities, and the mechanisms for existing learning.

This latter factor — the need to retain key staff — underscores many of the points made about the management of technological learning. It highlights the fact that technological learning is not only concerned with informed selection of projects, and their effective operational management, but it is also about incentives and the employment system, and includes organizational issues concerned with allowing employees to maximize their creativity. It is perhaps in this sphere that the lessons from "fast learning" firms such as Celltech are most apparent.

It is the management of the organizational and employment system around the knowledge base that allows firms to adopt learning-based, adaptive strategies. Such strategies may be appropriate for firms concerned with dealing with discontinuous technological change. As Loveridge and Pitt (1990) put it:

> ... effective strategies of technological innovation benefit from, indeed may necessitate an adaptive, learning orientation in which strategists and managers would do well to come to terms with the reality that they are rarely in full control of the firm's destiny. Thus adaptive, opportunistic knowledge appropriation as a means of effective learning is the key to technological transformation.

It might be the case, therefore, that the lessons learnt from detailed case studies of companies skilled at "fast learning" may be applicable to that large population of firms affected by turbulent technological change.

Notes

1   This chapter is based on Dodgson, M. *The Management of Technological Learning: Lessons from a Biotechnology Company*, de Gruyter, Berlin, 1991. Some sections describing the case study have previously been presented in Dodgson (1990a) and Dodgson and Rothwell (1990). The author would like to thank Ove Granstrand for his comments on a draft of this paper.
2   For a lay person's description of biotechnology, see Fairtlough (1984), Carey (1987), Sharp (1985).
3   For a full account of Celltech's history, achievements and failures, see Dodgson (1990a).

Appendix 1: Collaboration with universities and research institutes

Universities	Country	Research Institutes
19	UK	9
3	USA	3
2	Netherlands	1
1	Japan	1
1	Switzerland	
1	Belgium	
	Germany	1
	France	1

Appendix 2: Celltech's academic collaborations (current October 1989)

Research Collaborations	27 (34 personnel)
Co-operative Awards – SERC	3 (3 personnel)
– MRC	2 (2 personnel)
SERC CASE Studentships	27
Consultancies	45
SERC Clubs	3 (4 in 1990)
LINK Schemes	2 (4 in 1990)

References

Adler, P. (1988) "Technology Strategy: A Guide to the Literatures" Draft Report, Stanford University, December.

Ansoff, I. (1982) Managing Discontinuous Strategic Change: the Learning-Action Approach, in *Understanding and Managing Strategic Change*, (Eds I. Ansoff, A. Bosman and P. Storm), North-Holland, New York.

Argyris, C. and Schon, D. (1978) *Organizational Learning*, Addison-Wesley, Wokingham.

Arrow, K. (1962) The Economic Implications of Learning by Doing, *Review of Economic Studies*, June.

Arthur, B. (1989) Competing Technologies, Increasing Returns, and Lock-in by Historical Events, *The Economic Journal*, **99**(394), March.

Burns, T. and Stalker, G. (1961) *The Management of Innovation*, Tavistock, London.

Carey, N. (1987) Production and Use of Therapeutic Agents, *British Medical Journal*, 10 October.

Chandler, A. (1962) *Strategy and Structure*, McGraw-Hill, New York.

Cyert, R. and March, J. (1963) *A Behavioural Theory of the Firm*, Prentice Hall, Englewood Cliffs, NJ.

Dodgson, M. (1989) *Technology Strategy and the Firm: Management and Public Policy*, Longman, Harlow.

Dodgson, M. (1990a) *Celltech — the First Ten Years of a Biotechnology Company* Special Report, Science Policy Research Unit, University of Sussex, Sussex, UK.

Dodgson, M. (1990b) "Technological Learning, Technology Strategy and Competitive Pressures", paper presented at Process of Knowledge Accumulation and the Formulation of Technology Strategy workshop, Kalundborg, Denmark, May 20–3.

Dodgson, M. (1991) *The Management of Technological Learning: Lessons from a Biotechnology Company*, de Gruyter, Berlin.

Dodgson, M. and Rothwell, R. (1990) "Strategies for Technological Accumulation in Innovative Small and Medium-Sized Firms", paper presented at symposium on Growth and Development of Small High-Tech Businesses, Cranfield Institute of Technology, 2–3 April.

Dosi, G. (1982) Technological Paradigms and Technological Trajectories. A suggested Interpretation of the Determinants and Direction of Technological Change, *Research Policy* **11** (3).

Dosi, G. (1989) Sources, Procedures and Microeconomic Effects of Innovation, *Journal of Economic Literature*, **26**, September.

Fairtlough, G. (1984) Make Me a Molecule: the Technology of Recombinant DNA, *Proceedings of the Royal Institution*, 24 February.

Fairtlough, G. (1986) Creative Compartments, *London Business School Journal*, Summer.

Fildes, R. (1990) Strategic Challenges in Commercializing Biotechnology *California Management Review*, Spring.

Freeman, C. and Perez, C. (1988) Structural Crises of Adjustment: Business Cycles and Investment Behaviour, in *Technical Change and Economic Theory*, (Eds G. Dosi *et al.*) Pinter, London.

Hagedoorn, J. and Schankenraad, J. (1990) Inter-Firm Partnerships and Cooperative Strategies in Core Technologies, in *New Explorations in the Economics of Technical Change* (Eds C. Freeman and L. Soete), Pinter, London.

Hall, S. (1987) *Invisible Frontiers: The Race to Synthesize a Human Gene*, Sidgwick & Jackson, London.

Hambrick, D. and Crozier, L. (1985) Stumblers and Stars in the Management of Rapid Growth, *Journal of Business Venturing* **1**, 31–45.

Hamilton, W., Vila, J. and Dibner, M. (1990) Patterns of Strategic Choice in Emerging Firms: Positioning for Innovation in Biotechnology *California Management Review*, Spring.

Harris, R. and Mowery, D. (1990) Strategies for Innovation: An Overview *California Management Review*, Spring.

Hedberg, B. (1981) How Organizations Learn and Unlearn, in *Handbook of Organizational Design: Vol. 1* (Eds P. Nystrom and W. Starbuck) Oxford University Press, Oxford.

Horwitch, M. (1986) *Technology in the Modern Corporation: A Strategic Perspective*, Pergamon, New York.

Loveridge, R. and Pitt, M. (1990) *The Strategic Management of Technological Innovation*, John Wiley, Chichester.

Maidique, M. and Zirger, B. (1985) The New Product Learning Cycle, *Research Policy*, **14**.

Morgan, K. and Sayer, A. (1989) *Microcircuits of Capital*, Polity Press, Cambridge.

Nelson, R. and Winter, S. (1982) *An Evolutionary Theory of Economic Change*, Belknap Press, Cambridge, MA.

Oakely, R., Faulkner, W., Cooper, S. and Walsh, V. (1990) New Firms in the Biotechnology Industry: Their Contribution to Innovation and Growth, Pinter, London.

Orsenigo, L. (1989) *The Emergence of Biotechnology*, Pinter, London.

OTA (1988) US Investment in Biotechnology, *New Developments in Biotechnology*, **4**, Office of Technology Assessment, US Congress.

Patel, P. and Pavitt, K. (1988) The International Distribution and Determinants of Technological Activity, *Oxford Review of Economic Policy*, **4**(4).

Pavitt, K. (1987) Comment on Tushman, M. and Anderson, P. Technological Discontinuities and Organizational Environments, in *The Management of Strategic Change*, (Ed. A. Pettigrew) Blackwell, Oxford.

Pisano, G., Shan, W. and Teece, D. (1988) Joint Ventures and Collaboration in the Biotechnology Industry, in *International Collaborative Ventures in US Manufacturing*, (Ed. D. Mowery) Ballinger, Cambridge, MA.

Porter, M. (1985) *Competitive Advantage: Creating and Sustaining Superior Performance*, Free Press, New York.

Pucik, (1988) Strategic Alliances, Organizational Learning, and Competitive Advantage: the HRM Agenda, *Human Resource Management*, **27** (1).

Rosenberg, N. (1982) *Inside the Black Box: Technology and Economics*, Cambridge University Press, Cambridge.

Sahal, D. (1982) *Patterns of Technological Innovation*, Addison-Wesley, Reading, MA.

Senker, P. and Brady, T. (1989) Corporate Strategy: Skills, Education and Training, in *Technology Strategy and the Firm: Management and Public Policy* (Ed. M. Dodgson), Longman, Harlow.

Sharp, M. (1985) "The New Biotechnology: European Government in Search of a Strategy", Sussex European Papers, No. 15, Science Policy Research Unit, University of Sussex.

Snow, C. and Miles, R. (1983) The Role of Strategy in the Development of a General Theory of Organizations, in *Advances in Strategy*, (Ed. R. Lamb), JAI Press, New York.

Teece, D. (1987) *The Competitive Challenge: Strategies for Industrial Innovation and Renewal*, Ballinger, Cambridge, MA.

Tushman, M. and Anderson, P. (1987) "Technological Discontinuities and Organizational Environments" in *The Management of Strategic Change* (Ed. A. Pettigrew), Blackwell, Oxford.

Tushman, M., Virany, B. and Romanelli, E. (1986) Executive Succession, Strategic Reorientation, and Organizational Evolution, in *Technology in the Modern Corporation: A Strategic Perspective* (Ed. M. Horwitch), Pergamon, New York.

Yarrow, D. (1989) *US Biotechnology in 1988 — A Review of Current Trends*, Department of Trade and Industry, London.

7

CREATING DEMAND FOR BIOTECHNOLOGY: SHAPING TECHNOLOGIES AND MARKETS

Kenneth Green

INTRODUCTION

As explained in the previous chapter, biotechnology aims to exploit major scientific developments of the 1960s and 1970s, principally the technologies of recombinant DNA (rDNA, which made possible "genetic engineering", latterly "genetic modification") and of hybridoma production of monoclonal antibodies (MAbs). The first firms calling themselves biotechnology firms were set up in the late 1970s, launched on waves of venture capital in the USA and the UK. There was much optimism about the new products, particularly those for human healthcare, that biotechnology was to provide. However, the products have been slow a-coming. Whatever the radical potential of biotechnology's ability to manipulate genetic and cellular material, there has certainly been no applied biotechnology "revolution".[1] The OECD, in a recent review of biotechnology's impact, postpones the timing of any revolution to the next century, though it sees the 1990s as a decade in which many more biotechnology products will come on sale.[2] The large pharmaceutical companies have been buying up the smaller dedicated biotechnology firms since the mid-1980s, and have begun to boost their flagging profits with new ranges of genetically-engineered drugs. Genetically-engineered crops and foodstuffs are slowly being introduced into agriculture and supermarkets.

Biotechnology's under-achievement in the 1980s is not, perhaps, surprising. As Mark Dodgson explains elsewhere in his history of Britain's largest dedicated biotechnology firm, Celltech:

> [There] has been an underestimation of the time necessary for important scientific and technological discoveries with the potential to produce commercial and widely diffused new products to realize that potential. Semiconductors — a tangible *product* — invented in the late-1940s, took at least 25 years to develop into the range of applications we know today. The two major biotechnology breakthroughs of rDNA and MAbs — *process* inventions with no immediate

products — only occurred in the mid-1970s. It is unrealistic to have expected products to have emerged, and to have been widely diffused, in such a short period.[3]
    (Emphases in the original. He makes the same points in his chapter in this book.)

How does a new generic technology "realize [its] potential"? In the most obvious sense, it can do so by the formulation of products and processes based on an exploitative development of the core features of the new technology. Through the process of research and development, taking place in scientific research institutes and in the laboratories and workshops of firms, scientific understanding is "shaped" into specific artefacts. How this happens has been the subject of considerable research since the 1960s by economists interested in technological innovation and, since the 1970s, by sociologists interested in how technologies are socially constructed.

    There is, however, another aspect to the idea of a new technology "realizing its potential", an aspect which usually receives insufficient attention from economists and sociologists of technology alike. It concerns the "market" itself. Freeman defines innovation as: ". . . the first *commercial* transaction involving the new product, process, system or device",[4] (emphasis in original). For some new products the act of "commercialization" is their launch into an already established market, in which demand demonstrably existed before the new product was introduced. However, for the products of biotechnology, a new technology, the market in the sense of observable demand may not exist. The markets may have to "emerge" to go with the products. "Emergence", however, is a purely descriptive term. In addition to creating new artefacts, innovating firms may also be engaged in creating markets to go with them. Innovation in such circumstances — the commercialization of radically original products based on new technologies — has to involve the shaping of a market as well as the shaping of a technology; the shaping of the artefacts and the shaping of certain institutions that set up the demand for those artefacts must go together.

    In this chapter I shall explore what this simultaneous shaping of technology and market implies, using examples from the development of biotechnology in the 1980s. In particular, I want to examine the exploitation of monoclonal antibody (MAb) technology for the manufacture of new techniques for infectious disease diagnosis, and recombinant DNA technology for the production of veterinary products. These have proved to be the most successful commercial applications of generic strands of biotechnology so far. However, the innovating firms have always envisaged a much larger market for their products, and have sought to create new demand by a variety of means. In the event, such "market shaping" has proved more difficult than they had envisaged.

THE SOCIAL SHAPING OF TECHNOLOGY

Writers in the field of the social shaping of technology or the social construction of technological systems have been striving to provide credible models of how specific technological artefacts emerge as embodiments not just of internally-generated technical developments, but of broader social, economic and political forces. The most useful and influential collection of papers setting out the various approaches is in a volume edited by Bijker, Pinch and Hughes.[5] Three "schools" are identified — the "systems builders" (particularly Hughes), the "social constructivists", who apply ideas from relativist sociology of scientific knowledge to a new "sociology of technology" (Bijker and Pinch) and those working with the concept of the "actor-network" (Callon and Law).

MacKenzie has outlined three significant ideas used by Hughes in the latter's analysis of the growth of electricity supply systems from the late nineteenth century. Firstly,

[Hughes] shows that successful system builders *simultaneously* had to engineer technological matters (such as the design of a lamp filament), economic matters (such as the need to compete in price terms with existing gas suppliers), and political matters (such as the legislative frameworks within which electricity supply developed).[6]

Secondly, Hughes does not draw any sharp boundary between what goes on in a particular laboratory and what happens at the higher economic levels in a particular country. Economic (or it could be political or other social) obstacles to the building of some technological system become technical problems to be solved in the laboratory. Thirdly, and perhaps most intriguingly, system builders are not just passive *techniciens* constructing some artefact which then diffuses into a pre-existing socio-economic system external to the system-builder's world of the laboratory, having a subsequent "impact" on their environment; rather "they continuously seek to mould that environment so that the growth of the system is facilitated".[7] They act both technically and socially, and can thus be dubbed "engineer–sociologists" (as Callon puts it) or "heterogeneous engineers" (in Law's phrase).

Sociologists like Callon and Law have creatively elaborated such notions to give accounts of the actions of scientists and of technologists in the conduct of their work.[8] They go further than social constructivists like Bijker and Pinch, by refusing to give any privilege to the "social" sphere — a privilege which is contained in the very phrase "social shaping of technology": as if the social is all-powerful in the construction of artefacts, so that the rejected technological determinism is replaced by an equally totalitarian social determinism. Instead, Callon and Law focus on networks of scientific and engineer "actors" who work simultaneously on the technical and on the social worlds. For example, in their analysis of the development of the TSR2, a British military aircraft, Callon and Law describe how the RAF specified an aircraft in the early 1960s, in the face of Treasury, Ministry of Defence and Navy opposition. The specification thus had to take the form of more than a set of technical blueprints; it had to be a "socio-technical scenario", since it needed to include technical details of a desired weapons system/aircraft, together with assumptions about how "the political, bureaucratic and strategic world could be made to look five or ten years later".[9]

The crucial element of such analyses is to see the development of scientific knowledge and technological artefacts as an integral aspect of the simultaneous development of supporting social groups. In this sense, a technological artefact, in the sense of a concrete object, is one element of a "seamless web" of socio-technical relations. There is of course a physical difference between the artefact and the social arrangements of its use, but the artefact will be constructed with specific social arrangements running through it. In addition, it is important to remember that this process is not once-and-for-all. The actors need to be continually reproducing the "seamless web", by strategies of, in Callon's terms, *interessement* of existing and new *actants*, both human/ social and non-human/physical.[10]

When it comes to applying such ideas to the development of technological artefacts, the social constructivists and system-building analysts present problems for anyone interested in 1980s and 90s technologies. To begin with, the examples they analyse can be criticized for not being sufficiently contemporary. Both Bijker/Pinch and Hughes analyse artefacts or systems emerging in the late nineteenth/early twentieth centuries. Since then, however, technological developments have been carried out on a much larger scale, with different relations between interacting elements of the system and with strikingly different social groups being involved. In particular, technological development has its own professions and institutionalized forms; it is also "anticipatory" of demand (via forecasting techniques, for example) and a conscious constructor of "futures" through corporate strategies.

A more substantial failing is that they do not put technological developments in the context of one of the most significant organizational innovations of the last 100 years: the firm. Whatever the influences on innovation, whatever the nature of the actor-networks, the central institutions through which new products "emerge" are *firms*. It is from the organized activities of firms that the vast majority of new artefacts emerge, even if the research and development they do is paid for by others (such as in the defence sector). It is firms that innovate. Therefore it is *their* perceptions which are critical to technological shaping. *Pace* Constant, who sees technological knowledge as residing in an industry-wide network or in "communities" of industrial and university researchers, innovation as a socio-economic process of commercialization is centred on the actions of firms.[11] Anyway, even "networks" are now institutionalized increasingly through inter-firm agreements, in joint ventures and the like. Therefore it makes sense to view technological developments as being "refracted" through what goes on in firms. Whatever other actors do, it is firms' interpretations of what they do, mean, and want (or what they can be persuaded to do, mean and want) that provide the social influences on artefact production.

Economics of technological innovation

The notion that the activities of the *firm* are central to the contemporary innovation process can hardly be stressed enough. Any model of how technology is shaped or constructed has to focus on what firms do. Researchers in the economics of innovation of the last twenty years have made considerable strides in transcending the naive debates of the 1960s between those who saw innovation as being the result of the successful transfer of scientific knowledge ("science push") and those who saw it as the result of market demand ("demand pull"). More sophisticated analyses have seen innovation as the result of the coupling of market demand/user need with scientific/technical possibilities.[12] In the study of the diffusion of innovations, the naive idea of firms commercializing products then passively awaiting "demand" to develop, so that products diffuse through populations of "adopters", has also been superseded. Metcalfe has recast diffusion theory so that diffusion can be seen as an iterative process, between demand and supplier firms seeking to remain profitable as demand and competition change.[13] Metcalfe also emphasizes how initial ideas of level of demand are periodically revised as diffusion proceeds. Demand for any radically new innovation can thus only be *perceived* demand; market surveys and statistics of projected consumption are "images" or "visions" (or "forecasts") of future sales, to be revised as real purchases do or do not take place.

This still begs an important question, implicit in economists' distinction between radical, "science-based" innovations and incremental innovations, more subject to shifting demand patterns. Walsh presents a "dualist" model, in which new scientific ideas are essential in setting off a new technological paradigm; later when markets for the products of that paradigm's exploitation are established, subsequent innovations are more directly stimulated by changes in demand.[14] Schmooklerian demand-driven mechanisms predominate for incremental innovations, but one needs Schumpeterian concepts to explain how "new knowledge" establishes new industries based on radical innovations. Mowery and Rosenberg, in their reanalysis of a number of studies of the factors contributing to successful innovation have criticized the naive use of the concept of "demand" and "user-need" that some of the studies employ.[15] In their view, the economists' conception of demand only allows for incremental innovation, and therefore cannot account for the development of new demand, based on new technological exemplars themselves based on advances in scientific knowledge.

So how *is* demand established in the first place for radical innovations? How are markets actually created by firms for products for which no demand — in the strict economic sense — has existed before?

Sociologists have paid little attention to the social-institutional conditions for the existence and functioning of markets.[16] For the economist Moss, a market is "a set of social institutions in which a large number of commodity exchanges of a specific type *regularly* take place and to some extent are facilitated and structured by these institutions"[17] (emphasis added). To be somewhere for buying-selling transactions to take place on a regular or routine basis — in other words, for a market to exist — a framework of institutions has to exist or be established within which the sellers/buyers can act. "Institution" here is being used in the sense of including not just formally-constituted organizations — like manufacturing firms or retailers — but also socially-sanctioned arrangements and customs.

The institutions which have to be in place for a market to function are those in which prices are established, in which potential customers are informed that the goods are for sale, in which suppliers are informed there is a demand, and those whereby the goods and services are actually provided to the customer in transport and distribution systems; these are all continually subject to organizational innovation. In addition, regular exchange is dependent on legal institutions related to the exchange of property rights (a market is a "place" in which property rights to goods or services are exchanged, using money).[18] Normally these already exist and do not need recreating for new products. But sometimes they do need to change, particularly if the property rights had never been an issue before new technological interventions came up with new challenges to long-established rules, such as in the ownership of information (disputes over software reproduction) and of new plant/animal organisms.

Thus for new products, firms have to establish new institutions, or significantly modify existing ones, to establish a structure of routines within which regular exchange can take place. Before they can be sure of a market developing or growing, there needs to be institutional change which, in effect, constructs the "market-place" within which buying and selling can routinely take place. The institutional arrangements are a network which, to coin a new phrase, form the *market space* within which the marketplace works. For some products, even radical ones, changing the market space may be relatively easy — for example, some new electronic products can be sold through existing retailers to new market segments. For others, new arrangements have to be constructed for the market space to appear — and, crucially, innovating firms would have to be leading actors in this construction. Indeed the "radicalness" of an innovation might be defined in terms of the institutional changes — the amount of market space creation — that need to be effected before demand can "take-off" and the new product diffuse.

Commodification of technology

Writers on "commodification" and on "socio-political paradigms", drawing on work on the links between technological innovation and long waves of economic development, have made some contribution to the notion of how new demand for radical innovation arises.[19] They see new demand as resulting from changed human needs. The changes here are not, however, the result of changes in some imminent human desires or wants; they are rather the consequences of changed patterns of consumption, themselves induced by a change in a whole array of social, economic and political institutions. As the technological paradigm changes, so do institutions and so do the sorts of things people want. New technologies, new markets and new socio-economic arrangements all go together: consumption is restructured as part of social restructuring. Really radical new technological directions are thus associated with periods of intense social and economic change and crisis; new industries and new demand emerge.

Such analyses have been conducted at a high level of generalization. The setting of new product performance standards, deregulation, new forms of linkage between firms and between firms and public research institutions can all be seen, in the broad brush, as institutional changes required for the exploitation of new generic technologies. But what we lack is the detail of how this institutional change takes place, so that consumption desires change and radically new products are "wanted". In particular, we lack detail of how the product innovators — the firms from within which this demand is recognized — try to create such desires by altering the institutional framework of consumption.

Miles *et al.* have recently begun to explore these processes in detail (though not in the words I have used) with reference to the emergence of new consumer IT products of the late 1980s and 1990s — interactive compact disc systems (CD–I), interactive home systems, messaging systems (fax, email), drawing on previous work on home computers, HDTV and videotex.[20] They discuss how an innovating firm has, of course, to convince "itself" of the technical viability of any product; but the firms have also, they say, "to build a constituency behind the product". Building such a constituency will always involve convincing groups *within* the firm of the intended innovation's viability. External groups range from "selection environment" actors — such as regulatory agencies, standards-setting bodies, and external sources of finance — to other firms upstream and downstream in the production/selling chain — such as suppliers of complementary software/peripherals and distributors. Particularly crucial to constituency-building will be the conjuring up of desire among consumers for as yet non-existent products; IT consumer goods firms are especially active in projecting images of possible life-styles that imply the purchase of products that may be still only on the drawing-board. Such images are often encapsulated, claim Miles *et al.*, in the adjectives that innovators use to describe their products — "personal" communication networks, "home" automation. Market intelligence data on such putative demand can then be used to mobilize in-firm support for new product ranges.

This raises the question: how passive is the process of market development for the applications of new technologies? Having surmised that a new market could exist, and having the proto-product to satisfy it, a firm or industry will act to realize the potential demand. In markets where the potential user is knowledgeable (and where indeed the innovation may be a joint venture between user and producer) the acts might be relatively easy to identify; however, for the creation of mass markets, as the discussion by Miles *et al.* of IT consumer durables suggests, the process is likely to be very complicated. In the rest of this chapter I want to explore the process of market space creation in the production and sale of new products from biotechnology. The two examples I discuss are the use of monoclonal antibodies in new diagnostic tests for infectious diseases, and the use of rDNA technology to produce bovine hormones to increase milk production.

NEW MARKET SPACES FOR MONOCLONAL ANTIBODIES

Firms concentrating on development of the new biotechnology of the mid-1970s — based on explorations of the potential of recombinant DNA and/or of monoclonal antibodies — were almost all based in the USA or the UK. They have been called New Biotechnology Firms, or more suggestively, Dedicated Biotechnology Firms (DBFs), to indicate that they put all their resources into the exploitation of biotechnology's expected potential. At this time — from the late 1970s to the mid-1980s — those larger firms that might be expected to have been interested in rDNA and MAb technologies, such as chemical or pharmaceutical firms, were cautious. They were prepared to enter into joint ventures with the DBFs, but not to set up major biotechnology R and D, not to mention production and marketing programmes of their own. The DBFs were thus almost completely financed by joint venture deals with big companies and, to a lesser extent, by venture capital.[21] From the beginning, the DBFs tended to divide into those concentrating on exploiting rDNA technology (the prime example being Genentech) and those exploiting MAb technology (like Hybritech, Centocor, Monoclonal Antibodies and Genetic Systems) though others (like the UK's Celltech) were in both.[22]

Those exploiting MAb technology saw themselves in the long run as being in cancer therapeutics; MAbs were expected to be ideal "magic bullets" which could be biologically "designed" to zoom in on specific tumours, carrying with them appropriate anti-cancer agents. After all, much of the work leading to the 1970s advances in biotechnology was a direct offshoot of the huge US campaign to find cures for cancer that Nixon had initiated in the 1970s. However, there was pressure from the beginning to produce saleable applications of MAb technology, not least to demonstrate how exploitable the technology was and thus keep up the flow of investment cash into the firms. From the early 1980s therefore, the DBFs identified diagnostics as products that could be quick revenue-earners; research into therapeutic applications could be subsidised with the earnings (and further attractions of venture capital) from diagnostic products.

Diagnostic applications of MAbs would be easier to develop than therapeutic applications for three reasons. First, they could be made in low volumes, so scale-up problems from laboratory methods would be few. Second, extensive in-human testing would not be required, since MAb diagnostics were intended for *in vitro* use. This was a very important factor; therapeutics would be subject to the stringent regime of animal testing and clinical trials that are required to get government approval for human *in vivo* use for any drug-type product, whatever the technology on which it is based. Third, MAbs could be used for quick, cheap diagnosis for a range of infectious diseases for which no equivalent, competing technology existed; therefore completely new markets were thought to exist for these first major applications of the new technology. As it turned out, this was no more than an assumption — as, of course, most market predictions are — but an assumption that proved to be the most weakly based of the reasons for going into MAb diagnostics.

Before exploring the market question further, it is worth pointing out the extent to which the particular direction of MAb technology exploitation was governed by the institutional forms within which the exploitation took place. The shaping of MAb technology in its diagnostic direction is obviously related to the financial pressures to come up with saleable products sooner rather than later. In addition, the particular regimes of regulation in force in the USA and UK, as enforced by the pharmaceutical and (in the US) medical device approval schemes, determined which particular MAbs would be developed. One could view this in terms of evolutionary economics — technological development being pushed along a particular trajectory by a specific financial-regulatory selection environment. However, following the arguments presented earlier, one would also have to explore how active the MAb DBFs were in creating the market for their products.

Diagnostics using antibodies to detect the presence, in blood, urine and other bodily fluids, of biological molecules (like proteins) or infectious agents (like bacteria or viruses) are called immunodiagnostics. Immunodiagnostics have been developed from the 1960s using antibodies to human antigens produced by injection into animals. Antibodies produced thus tend to be complex mixtures of different types of individual antibodies, and are called polyclonal antibodies; monoclonal antibodies, produced by 1970s' developments in cell culture and screening techniques, were thus an advance in that they were "pure" antibodies, specific to one antigen each. Polyclonals could only be produced for a limited range of antigens, and then only for biologically-active proteins. In theory, monoclonals could be produced specific to any biologically-active molecule, so that even diagnostics for bacteria and virus identification by microbiologists would be affected.

By the early 1980s the market for immunodiagnostics for use in medical laboratories was well established in the USA and Europe. The early 1980s also saw the beginnings of retrenchment in the US (and European) healthcare system. Cutbacks in healthcare spending by the Reagan government slowed the growth of the immunodiagnostics markets which Abbott, the leading US laboratory instrumentation and reagent manufacturer, had developed. Abbott still managed to prosper, based on its policy of giving away its analysers. Other firms, notably Abbott's main rival, Syva, did not. All diagnostics firms began to explore two other markets that were thought to have some potential.

The first was the so-called market for Physician's Office Testing (POT), testing done not by a hospital-based or private laboratory, but in the office/surgery of general practitioners. In the USA (but not in all European countries) the cost of this testing is reimbursed by health insurance agencies. It was thus seen as a potentially lucrative market, bypassing the financially-constrained hospital laboratory sector. Physicians could increase their income by performing the tests themselves (or rather, setting up a small in-office laboratory and hiring technicians or nurses to do it). However, the analytical techniques and the equipment they used would have to be simpler than that used in hospital laboratories, given the lower level of skill that would be expected of physicians, nurses and unsupervised technicians. The requirement was thus for quick, easy-to-perform, low-skill assays.[23]

The second market was, potentially, even larger than the POT market. If the tests/assays were easy enough for untrained doctors to perform, they might be easy enough for the patients themselves. The tests could then be bought in chemists/pharmacists shops in a DIY form, as over-the-counter (OTC) tests. Given the range of infectious diseases that could be tested for, the potential market might be huge. For example, patients could cease to rely entirely on the formal medical system to diagnose the presence of sexually transmitted disease (STD) infections. And then there was the possibility of DIY screening for cancers . . . .

The diagnostic applications of MAb technology that the DBFs were working on from the late 1970s were emerging in the context of the market structure and promise just described. By 1985, biotechnology-based products were overwhelmingly MAb-based: 100 firms were doing MAb R and D, and 600 MAbs were available; 36 per cent of DBFs were in diagnostics; sixty diagnostic kits had been commercialized.[24] However, DBFs had limited success in entering diagnostics markets on their own. Not surprisingly, there was reluctance to take on firms like Abbott in the well-established hospital laboratory immunodiagnostics market.

Another DBF, Genetic Systems, attempted to develop MAbs to make up into kits for STD detection, for sale to hospital laboratories, physicians' offices and, it was surmised, eventually as OTC products. STD tests were identified as a potentially rapidly-growing market, given the fact that STDs were growing fast in the population (even before the discovery of HIV-related conditions), and the existing technologies were slow and required the skills of a microbiology laboratory.

However, Genetic Systems, among others, failed to develop this market. By 1985 no firm, DBF or large diagnostic manufacturer, was making money from infectious disease diagnostics. As Teitelmann puts it, "for all the studies and all the forecasts, the microbiology market refused to accept new products in large quantities ... [because of] a fatal combination of technical inadequacies and marketing misjudgements."[25] Teitelmann analysed the failings thus:

1   To perform an STD test requires taking a swab of the infection — for a woman from the vaginal canal, the cervix and the anus, for a man from the urethra and the tip of the penis; this is a difficult procedure even for a doctor to carry out, and it is unlikely that it could or would be performed correctly by an untrained and squeamish "patient" for a DIY test. Any faster, more accurate test, even if based on MAb technology, is thus not much of an advance, if the swab-taking is as difficult as ever it was.
2   The concept of "the customer" for STD tests was ill-defined; Abbott's immunodiagnostics market was built as a complement to its clinical chemistry sales; however, STD tests for laboratories were aimed at microbiology laboratories, a smaller and traditionally more conservative group.
3   In the 1970s, the most popular immunoassays were based on radioactive detection (RIAs); because microbiologists were not skilled in the use of radioactive chemicals, clinical chemists, who *were* skilled, were the obvious target for RIA polyclonal products; however, the 1980s saw the coupling of MAb technology with labelling based on enzyme reactions (so-called enzyme-linked immunosorbent assays, ELISA); in principle, this made the tests easier to perform, but it also sharpened the demarcation dispute between clinical chemists and microbiologists; RIA-based tests have the advantage, as far as a clinical chemist is concerned, of being quantifiable; ELISA-based tests are not, yet quantifiable tests

are important if antibiotic therapies are to be monitored. So the new tests could satisfy neither the clinical chemists nor the microbiologists. Ironically, Abbott side-stepped this dispute by going straight to the less professionally-disputatious physicians' office market, selling tests for gonorrhoea and strep throat.

In short, Genetic System's failure to develop a market for MAb-based STD tests is a good negative example of the notion that an identification of the "market" is sufficient. STDs were indeed increasing in incidence, and there might thus have been a "need" to identify which infection someone has, so as to get the right treatment quickly. But the conclusion that new, STD-specific tests were therefore bound to sell, does not mean much in the absence of an appropriate market institutional structure. In this case, to create a market space would have required the innovator, in Callon's formulation, to "interest" a number of *actants.* the core MAb technology being only one; a further actant is the socio-technical system of sample collection. The human actants are the laboratories themselves, subject to professional demarcations and habits of working. In addition there are the "selection environment" constraints of the regulatory system, which required approval of new medical devices. In the early 1980s struggle to commercialize MAbs, it turned out that the DBFs were unable to "interest" the necessary actants. The larger firms, who had already established a market space for hospital laboratory testing and were carving one out for POT, were able therefore to develop markets, to commercialize MAb innovation. However, all firms have found difficulties in creating the market space for the potentially largest market of all for MAb diagnostics, for OTC products.

The creation of new markets for MAb-based diagnostics has not fulfilled the expectations of early 1980s biotechnology enthusiasts. DBFs found it relatively easy to sell their MAbs, either built into their own kits or, more often, as components of kits made and sold by established immunodiagnostics firms to medical laboratories. MAbs were, in effect, just incremental (if important) innovations within immunodiagnostic markets. However, more radical applications of MAbs to STD and for POT proved more difficult. The market spaces for such tests required an engineering of institutional arrangements, as well as appropriate technological developments. The study of the development of the pregnancy test market, into which MAbs slotted in the mid-1980s, shows what institutional engineering was necessary.[26] In this case, innovating firms had to negotiate with regulatory agencies (including, in the USA, the legal system) and, on a continuing basis, had to convince and placate professional medical practitioners hostile to over-the-counter testing.

Therefore, any substantial success for MAbs would have required assaults on a number of fronts. It is hardly surprising that the DBFs, even in league with bigger firms more accustomed to traditional markets, found the creation of large market spaces for their products exceedingly difficult.

NEW MARKET SPACES FOR GENETICALLY-ENGINEERED HORMONES[27]

Bovine Somatotropin (BST) is a hormone discovered in the 1930s. Amongst other things, it controls the conversion in cows of animal food into milk. It was soon demonstrated that its administration, by injection or implantation, increased production of milk for a given quantity of food. The isolation of protein molecules like BST from bovine tissue used to require the laborious processing of thousands of pounds of animal glands. The advent of recombinant DNA technology in the mid-1970s, however, opened up the possibility of producing such molecules in hitherto inconceivable quantities at a price that would command enormous sales, to the tune of $1 billion per annum.

This mass-produced BST is not itself a genetically-engineered product; rather it is made available by a process that relies on genetic engineering. (Indeed, the BST produced by this process is not chemically identical to "natural" BST — so it has come to be known as recombinant BST, rBST.) rBST's radical nature therefore can apparently be sought in its sudden availability at a low price. But this sudden shift in the supply curve did not, as it turned out, result in a straightforward surge in demand. As rBST's innovators have discovered, a new market space had to be created for the expected demand to be realized. rBST's radicalness may be seen in the need for "institutional engineering" to establish it as a routinely marketable product.

By the early 1980s a group of American veterinary product companies led by Monsanto (the others being American Cyanamid, Eli Lilly and Upjohn), were gearing up for a major sales drive of this new product, by organizing tests on farm animals to which rBST was administered on a trial basis. The purpose was to confirm the finding of increased productivity and to obtain regulatory approval by satisfying the appropriate authorities, in the USA and the EC, that no residues harmful to humans would appear in the milk or the meat, and that no harm would come to animals reared in this way. The UK was selected as the test country in which marketing approval was to be sought first, as a regulatory bridgehead that would, it was hoped, lead to approval throughout the EC. However, rBST gradually began to encounter more and more criticism. Indeed the choice of name, "bovine somatotropin", was itself, it has been alleged, the result of adverse publicity, which led companies to de-emphasize the fact that it was a hormone, lest public acceptance be further prejudiced. (In the 1930s it had been named bovine growth hormone, BGH; the genetically-engineered homologue is sometimes called rBGH as well as rBST.) The tangled issue of regulating the use of hormones in agriculture continued throughout the 1980s to produce shifting responses from the regulators, with some uncertainty among the public as a result, about which hormones should rightly be considered hazardous and which not, regardless of the scientific evidence.

The criticism focused on three issues. First there were those who claimed that sufficient amounts could be ingested by drinking milk to cause harm to babies and children. Second there were those who claimed that cattle treated in this way were in fact subject to veterinary problems, and the use of the product was therefore inhumane. Third it was clearly hard to deny that rBST would be likely to exacerbate an already severe problem of over-production of milk, which in 1986 led governments across the EC to implement cuts in milk production quotas and to take other measures, equally problematic politically, to encourage farmers to produce less milk. One could argue that in the longer term, any measure that allowed producers to be more efficient was advantageous to them, even in a saturated market. But such measures would still require farmers to reduce their herd size and to farm with more scientific control over their activities. Many small farmers felt that this was beyond their means.

The controversy continued throughout 1990. In the USA, milk from rBST-treated cows is now on sale; the National Institute of Health has judged the milk safe for human consumption, though many US supermarkets refuse to sell it due to consumer objection. The US veterinary product regulatory authority, the FDA, has yet to pronounce on the safety of the product for cows. The EC has not yet given its verdict, and the UK's Veterinary Products Committee has refused to approve Monsanto's version of the hormone, so it cannot be sold in Britain. It is also banned in Norway, Sweden and Denmark. Approval for use in the USSR and Czechoslovakia has, however, been granted, and a development agreement between Eli Lilly and the Indian National Dairy Board has been reached.

Viewed from within the rBST manufacturers' camp (organized through their trade association, whimsically named the National

Organization of Animal Health, NOAH) the creation of a market space for their product has, so far, proved remarkably and surprisingly difficult. Surprising for two reasons: first, the use of new biotechnologies to make products for *animals* rather than for humans was identified in the early 1980s as a fast track to success, since it avoided the much tougher regulatory systems required for human pharmaceuticals (similar reasons were given for the use of MAbs in human diagnostics rather than therapetics); second, rBST, like human insulin produced by similar manufacturing methods, was deemed identical to a "natural" hormone and unlikely, it was assumed, to have any significant side-effects. It seems probable that such a product, if introduced in the 1960s (or earlier), would have had little difficulty in obtaining regulatory and agricultural approval. However, the institutional arrangements and array of *actants* in the 1980s have been such that Monsanto and NOAH have been presented with a difficult task.

There have been a variety of *actants* opposed to the use of rBST. A conference to discuss BST in 1988 heard papers from a Monsanto UK executive, a German Green Euro-MP, veterinary and animal welfare experts, and a consumer lobbyist from the London Food Commission.[28] The Commission's BST Working Party had representatives of a wide range of environmentalist, animal welfare and consumer activist bodies, as well as women's organizations and agricultural workers. These groups, as well as farmers' organizations, have lobbied Agricultural Ministries in EC countries and have been supported by pro-small-farmer and pro-green parties in the European Parliament. Any technological development that can be shown to have implications for the Common Agricultural Policy is now bound to involve the European Parliament and the European Commission.

The response of Monsanto and the other companies has taken place on a number of fronts. As is common with disputes over the safety and efficacy of pharmaceutical substances, the innovating firms have turned up the scientific power — in this case, 130 studies have been carried out to demonstrate rBST's safety to cows and people. In addition there have been numerous studies of rBST's economic efficacy for farmers and, through market surveys, of consumer opinion. By these means, the innovators have sought the support of mainstream scientific opinion and regulatory authorities. It is reported that the pile of papers submitted to the regulatory authorities is 70 feet high. Monsanto has also taken the novel approach of establishing "focus groups", which are "neighbourhood workshops, organized to exchange views on BST with farmers, veterinarians, doctors, teachers, nurses and consumers", in which to press their case for the use of rBST.[29]

Such action in favour of market space construction has so far faltered in the historically novel politics of the 1980s and 90s — new politics of environmental concerns, animal rights, consumer activism and, in Europe, agricultural surpluses and the powers of the European Commission and Parliament. The debate about BST has come to be regarded as a struggle between competing ideas about "natural" limits of agricultural animals and the value of different farming practices. The introduction of BST has run into difficulties because its proponents have failed so far to build sufficient support for this elaboration of intensive dairy farming. Thus the proponents espouse a form of farming based on the constant use of science to create and rear even more productive animals. For them, the role of biological science is to evade or reset any natural limits to productivity. For the opponents of BST, that is precisely the problem: that scientific means are being used in pursuit of economically inappropriate and morally indefensible ends, in overriding these natural limits.

CONCLUSION

Similar problems in the creation of markets for other, new biotechnology-based products can be expected in the 1990s. Now under development as a possible sideline for farmers, transgenic animal technology, notably the production of high-value biologicals such as clotting factor, from genetically-modified sheep, is one such example.[30] The two examples of attempted market creation for new biotechnology-based products discussed in this paper are notable principally because they were originally considered fairly non-controversial — they did not involve the genetic modification of higher organisms and, crucially, they did not raise regulatory problems of a scale that might be anticipated for any new human therapeutics.[31] Yet the creation of a market has proved extraordinarily difficult. Shaping biotechnology into saleable products thus seems to raise a bigger agenda of industrial and social institutional "restructuring" issues than the simple products it has sought to sell so far might suggest.

Of course, this is partly due to the fact that biotechnology is irredeemably controversial. It raises questions about what is "natural" and, given contemporary "green" political concerns, it invites the mobilization of a wide range of social groups with their own views on "nature" and how human societies should interact with it. Market creation in such circumstances is bound to involve innovating firms in novel "constituency-building", as much as in novel technical inventiveness. However, as Miles *et al.* have shown, similar considerations apply to other generic technologies. Turning information technologies into new consumer products that utilize what is new about IT, namely its reconceptualization of the handling of information (rather than just making higher-tech versions of consumer durables first commercialized before 1980), is almost as difficult as commercializing biotechnology.

Radical innovation as the creation of market spaces has conceptual relevance, therefore, to the exploitation of all generic technologies. As this chapter has pointed out, while focusing attention on the firm as the modern institutional driving force of innovation, writers on the economics of innovation have paid little attention to the ways in which markets are actually created, in terms of the strategies which are adopted to alter institutional arrangements. Writers on the social shaping of technology should have much to offer on this theme of how firms seek to mobilize support and engage in "political" action in pursuit of commercialization.

Unfortunately sociologists of technology, while providing valuable insights into how social (economic, political and ideological) factors push technologies into particular lines of development, have not paid much attention to the firm as the institutional locus of "technology-shaping". Social factors, whether economic signals or political influences, do not "enter" the site of product/artefact development — the R and D department — directly. They are perceived, collected and interpreted by other departments of the firm whose members interact with R and D. As Vergragt points out in his analysis of the development of flame retardants at AKZO, the actors involved in technological shaping are not confined to R and D technical people, but are also those engaged in relations with the outside market in the marketing/sales department and in senior management.[32] It is the interaction between the shaping of technologies and the shaping of markets that constitutes innovation; it is an interaction that deserves greater attention from sociologists and economists of technology.

Notes and references

1   For a survey of biotechnology's "revolutionary" promise see Marx, J.L. (ed.) (1989) *A Revolution in Biotechnology*, Cambridge University Press, Cambridge.

2   Organization for Economic Co-operation and Development (1989), *Biotechnology: Economic and Wider Impacts*, OECD, Paris, chapter 3.

3   Dodgson, M. (1990) *Celltech: the first ten years of a biotechnology company*, Science Policy Research Unit Discussion Paper Series, Brighton, 86.

4   Freeman, C. (1982) *The Economics of Industrial Innovation* (Second Edition), Frances Pinter, London, 7.

5   Bijker, W.E., Hughes, T.P. and Pinch, T. (eds) (1987) *The Social Construction of Technological Systems: New Directions in the Sociology and History of Technology*, MIT Press, Cambridge, MA.

6   Mackenzie, D., "Missile Accuracy: a case study in the social processes of technological change," in Bijker *et al.*, op. cit., 196.

7   Ibid., 197.

8   See Callon, M. "Society in the Making; the study of technology as a tool for sociological analysis," and Law, J., "Technology and Heterogeneous Engineering: The case of Portuguese Expansion", in Bijker *et al.*, op. cit.

9   Law, J. and Callon, M. (1988) Engineering and sociology in a military aircraft project: a network analysis of technological change, *Social Problems*, **35** (3), 284–90.

10   The concepts of *interessement* and *actant* are used by Callon in his analysis of a piece of scientific research (see Callon, M., "Some elements of a sociology of translation: domestication of scallops and the fishermen of St Brieuc Bay," in J. Law, (ed.) (1986) *Power, Action and Belief: A New Sociology of Knowledge*, Routledge & Kegan Paul, London). To borrow his terminology for this case: an innovating firm's *problematization* of its innovating mission requires it to enrol/interest not only the technical developments of the artefact, but also those *actants* in the socio-economic world who are required to change to open up the market space for the innovation. Consumers are of course the ultimate *actant* (since they buy the product) but the space within which they buy is structured by the market institutions, which the innovating firm has to get redesigned. The crucial point is that the R and D–marketing interaction is the central actor here, roping in (enrolling, *interessement*) the social and technical worlds to crystallize them in specific products.

11   Constant, E., "The Social Locus of Technological Practice: Community, System, or Organisation," in Bijker *et al.*, op. cit.

12   See Freeman, C., op. cit., chapter 5.

13   Metcalfe, J.S. (1981) Impulse and Diffusion in the Study of Technical Change, *Futures*, **13**, 347.

14   Walsh, V. (1984) Invention and innovation in the chemical industry: Demand–pull or discovery–push? *Research Policy*, **13**, 211.

15   Mowery, D. and Rosenberg, N. (1979) The influence of market demand upon innovation: a critical review of some recent empirical studies, *Research Policy*, **8**, 102.

16   An exception is Burns, T.R. and Flamm, H., (1987) *The Shaping of Social Organization*, Sage, London; however, Chapter 8, "The structuring of markets" concludes: "The analyses of this chapter have ignored, among other things, *innovations in production, products and marketing* which may facilitate the break-down or erosion of market exclusion rules." (p. 150).

17 Moss, S.J. (1981) *An Economic Theory of Business Strategy*, Martin Robinson, London, 1–2.

18 Hodgson, G.M. (1988) *Economics and Institutions: A Manifesto for a Modern Institutional Economics*, Polity Press, Cambridge, chapter 8.

19 Gershuny, J. and Miles, I. (1983) *The New Service Economy: the Transformation of Employment in Industrial Societies*, Frances Pinter, London; Blackburn, P., Coombs, R. and Green, K. (1985) *Technology, Economic Growth and the Labour Process*, Macmillan, Basingstoke; Perez, C. (1983) Structural Change and the Assimilation of New Technologies in the Economic and Social System, *Futures*, **15**, 357.

20 Miles, I., Cawson, A. and Haddon, L. (1990) "The Shape of Things to Consume", paper for PICT Workshop on Consumption, University of Sussex, mimeo; see also, Miles, I. (1988) *Home Informatics: IT and the Transformation of Everyday Life*, Pinter, London.

21 For data on the growth of biotechnology firms, see Orsenigo, L. (1989), *The Emergence of Biotechnology: Institutions and markets in Industrial Innovation* Pinter, London; Sharp, M. (1990) "Technological Trajectories and Corporate Strategy in Biotechnology", University of Sussex, mimeo, 1990; Kenney, M. (1986), *Biotechnology: the University–Industrial Complex* Yale University Press, New Haven, CT.

22 See Orsenigo, op. cit., chapter 3.

23 For details of the kinds of tests physicians were being recommended to perform, see Fischer, P.M., Addison, L.A., Curtis, P. and Mitchell, J.M. (1983), *The Office Laboratory*, Appleton-Century-Crofts, Norwalk, OH, and Baisden, C.R. (1985), *The Office Laboratory*, Aspen Systems Corp., Rockville.

24 See Orsenigo, op. cit., chapter 5.

25 See Teitelmann, R. (1989), *Gene Dreams* Basic Books, New York, 156.

26 See Green, K. and Hyde, B. (1989), "Near-Patient Testing: An Annotated Bibliography", mimeo, Manchester School of Management, UMIST, section 3.

27 A shorter version of this section has appeared in Green, K. and Yoxen, E. (1990), The Greening of European Industry: What role for biotechnology?, *Futures*, June, 475–95; see also Wheale, P. and McNalley, R. (eds) (1990) *The Bio-Revolution: Cornucopia or Pandora's Box?* Pluto Press, London, and articles in *Science*, **249**, 1990, pp. 852 *et seq.*

28 Brunner, E.J. (1988), *Bovine Somatotropin: a Product in Search of a Market* London Food Commission, London.

29 See article by Robert Deakin of Monsanto, in Wheale and McNalley (eds), op. cit., 72.

30 Human proteins, when made in bacteria, tend to lack the attached sugar molecules that confer functionality, and to be very hard to extract from bacterial cultures. It has been possible for several years to insert gene sequences into animal embryos at a very early stage, such that the cells of the adult animal make

large quantities of the protein specified by the inserted gene. The problem is generally one of control, but in recent years greater success has been achieved in controlled synthesis of valuable human proteins into the milk of sheep and cattle. Two problems hang over the potential exploitation of this technology by farmers. One is the scale of demand, which would probably not support very many farmers, the other is the controversial nature of using animals in this way, as attitudes to animals change in some European countries.

31   Nevertheless, therapeutics have also had their problems. For a discussion of Genentech's innovative failure with tissue plasminogen activator (tPA) see Westphal, C. and Glied, S., (1990), AZT and tPA: the disparate fates of two biotechnological innovations and their producers, *Columbia Journal of World Business*, Spring/Summer, 83–100. In Germany, the commercial production by genetically-modified organisms of something as apparently uncontroversial as insulin has been prevented by persistent anti-biotechnology campaigning by environmentalist organizations.

32   Vergragt, P.J. (1988), The Social Shaping of Innovations, *Social Studies of Science*, **18**, 483.

8

THE POLICY RELEVANCE OF THE QUASI-EVOLUTIONARY MODEL: THE CASE OF STIMULATING CLEAN TECHNOLOGIES

Johan Schot

INTRODUCTION

In mainstream neoclassical economic thinking, technology is an *exogenous* factor: the moment a specific technology is required, it can simply be taken down from the shelf. Neo-Schumpeterian economists object to this view. They wish to make technological development an *endogenous* part of their models. This implies that the interaction between technological and economic development must be taken into account. The model must offer sufficient scope for the continuous change and mutual adjustment of the economy and technology. (For a recent overview of Neo-Schumpeterian economics, see Dosi *et al.*, 1988.)

In the sociology and history of technology, attempts are made to unravel the interaction between technological and societal development. Discussion in these disciplines has focused on dichotomies: internal versus external approach, and/or content versus context.[1] Until recently, very little progress had been made in this discussion. Many analysts simply declare internal and external factors complementary to one another, without clarifying how that complementarity, and thus interaction, should be conceptualized. Empirical studies often contain a detailed description of technological development, supplemented with a list of those factors that had an influence on this development. As Law observes (1987a, 411): ". . . one is presented with a balance sheet with society (or the economy, or science, or politics) on the one hand and technology on the other. Analysis becomes the study of transfers between columns . . .". Recently, however, several general models have been developed within the disciplines of sociology and the history of technology which focus on the modelling of integration between content and context — for instance, the systems approach (Hughes, 1983; Staudenmaier, 1985), social constructivism (Pinch and Bijker, 1987) and the actor network approach (Latour, 1987). These approaches do not differentiate a priori between content and context. Their close interrelation is referred to in this approach by the metaphor of the seamless web.

It is not my intention to discuss these economic, sociological and historical models at full length. The angle of approach taken is this: which points of departure do the Neo-Schumpeterian approach, social constructivism, the actor network and system approach offer to help identify possibilities to steer the development of technology in societally desirable directions? This effort is called (at least in the Netherlands) constructive technology assessment (Rip and Van den Belt, 1986; Schot, in press).

This chapter first analyses Neo-Schumpeterian economics and sociological and historical models of technical change. Neo-Schumpeterian analysts have proposed an evolutionary model of technical change — so there is variation and selection, the basic building blocks of evolution. Sociologists of technology strongly oppose an evolutionary model. They deny the existence of an independent selection-environment, and stress that the content of technologies is shaped simultaneously with its context. This implies that one should rather conceptualize technical change in terms of a co-evolutionary development. This chapter will compare and evaluate the different approaches, and propose a quasi-evolutionary model which focuses on the way variation and selection are linked by actors, for instance firms. The notion of *technological nexus* as a link between variation and selection will be put forward.

The chapter goes on to discuss the possibilities of steering technical change, using the promotion of clean technologies as an example. On the basis of the quasi-evolutionary model three basic strategies will be delineated and discussed.

The chapter finally outlines a new role for the government in the "technology development game", and draws some general conclusions.

EVOLUTIONARY, CO-EVOLUTIONARY AND QUASI-EVOLUTIONARY MODELS OF TECHNICAL CHANGE

The relation between variation and selection

Neo-Schumpeterian economists conceptualize technological change as evolutionary. Technological development is a search process involving trial and error and uncertainty, and various options are implemented and evaluated. It can be conceptualized as a sequence of variation and selection processes. Variations are by no means tried out purely at random during the search process; heuristics are deployed. Heuristics as guidelines (rules of thumb) which promise, but do not guarantee, finding solutions to problems. Employing heuristics reduces the uncertainty inherent to the search process. The presence of heuristics implies that technological developments follow quite specific directions, while other possible directions are ignored.[2]

Selection between variations occurs in two ways: *ex ante* and *ex post.* *Ex post* selection obtains when the products and processes produced by heuristics are exposed to market selection pressures. The Social-Darwinist principle of survival of the fittest plays a major role here. Variation is stochastic or random; in other words, generation is independent of selection. *Ex ante* selection implies that influence is exerted on the generation of variations, and thus on the shaping and the choice of heuristics. This form of selection takes place when firms *anticipate* possible selection by the market. Neo-Schumpeterian economists use the term "selection environment", rather than the term "market". The term selection environment covers not only the neo-classic concept of market (structure and size of supply and demand, prices) but also various institutional factors (rules, relations between employers and employees, political structure) and geographical factors (Dosi, 1988).

Analysts using the actor-network and system approaches typically deny the presence of independent selection. These analysts begin by refusing to distinguish rigorously between the social (including economic) and the technological realms. They fiercely oppose a dual repertoire using different concepts for analysing the content of technological development and the influence of the surrounding environment on this technological development. It is emphasized that the content of technological development is shaped simultaneously with the context. Each variation has a script, or scenario (Akrich, 1988) which, in addition to the technical aspects, also includes aspects of the surrounding environment. Callon (1986), for example, showed how Electricité de France engineers, early in the 1970s, simultaneously designed not only an electrically driven car, but also the total environment in which that car should function. Government regulation, the research programmes and the production methods of car manufacturers, all had to be changed. This meant that engineers would be involved not only with the technical aspects (batteries, fuel cells, etc.) but also with the economic (cost, financing) and political aspects, precisely those elements referred to as the selection-environment in the Neo-Schumpeterian approach. For this reason those who develop technology need to be "heterogeneous engineers" (Law, 1987b). They are not only responsible for the technical realization of their design, but also for the social side. In other words, they have to recruit a whole range of *heterogeneous* elements in order to realize their design. In short, these analysts made no mention of independent variation and selection, the materials of the evolutionary model, but rather stress the co-evolution of both technology and selection environment.[3]

This yields a significant advantage over the economic model, in which little attention is given to the influence of the variation process on the selection environment. Sociological and historical studies have shown that social, political and economic factors are embedded, as it were, in technological development. Successful variations, therefore, also change the selection environment. Contrary to the Neo-Schumpeterian approach, therefore, more justice must be done to the influence exerted by the variation process on the selection process. In particular, the actor-network denies wrongly any independence of the selection environment. Certain contextual factors do quite definitely exert an influence on the content of scripts that are orchestrated and on the further realization of those scripts, without their being actually changed. There is, therefore, a "hard, structured selection environment" which exerts an influence on variations which manifest themselves by blocking certain variations and encouraging others in their further development.[4]

There are three ways to distinguish how the variation and selection processes are linked by actors. First of all, actors (firms, for instance) may anticipate the later selection from the selection environment by adjusting the heuristics. This is what Dosi referred to as *ex ante* selection. Second, institutional links may be created between the variation process and the selection process. These links help translate certain environmental requirements into criteria and specifications used for the development of technology. I would like to introduce the term *technological nexus* for these links (for a more extensive elaboration see Schot, in press).

Third, actors might attempt to create a niche in order to protect variation (and expectations in respect of variation) against selection which could be too rapid and drastic. Rip (1989) identifies this phenomenon with the term "strategic niche management". This niche can be diminished in stages. Firms do this by phased research development via tests, upgrading of tests, trial production and production.

In short, combining the valuable aspects of the evolutionary and co-evolutionary models, we might, following Rip (1989), devise what could be called the quasi-evolutionary model. In this combined model variation and selection are neither independent nor coincidental processes. Selection may be either anticipated or temporarily excluded or attenuated in the variation process. Furthermore, institutional links exist between variation and selection; as we shall demonstrate, these technological nexus offer powerful possibilities for directing technological development.

The role of actors in technological development

The quasi-evolutionary model shows that we can distinguish analytically between three sorts of actor:

1 Actors who are directly involved in the formulation of objectives and heuristics. By doing this they determine the content of variation generation. Good examples of this are company R and D departments, and technological institutes funded by the government.

2 Actors who attempt to selectively influence the variations from outside in order to obtain desired effects. These are actors who do not formulate the objectives of technology development themselves. Good examples are government bodies that try to force technological change by way of environmental regulations, and environmentalists, who attempt to do the same by campaigning.

3 Actors who couple variation and selection: the technological nexus. These are departments or individuals who, on the one hand, translate demands made from within the selection environment into recommendations or objectives for technological development, and who impose the demands made by certain technological variations on the selection environment. Examples are company environmental departments. Such departments have to translate environmental regulations in technological specifications, but also attempt to influence those regulations in such a way that they are congruent with the demands set by the technology which is used by their company.

STRATEGIES FOR STIMULATING CLEAN TECHNOLOGIES

The implementation of environmental policy in industrialized countries has led mainly to innovations in "end-of-pipe technology". Innovations in "clean technologies" are lagging behind. The essence of the end-of-pipe technology approach is to treat the residuals the production process generates, but to leave the production process itself unchanged. The residuals are modified, so they are less noxious, easier to store, or reusable. An extreme category is technology to deal with environmental damage done in the past. By contrast, clean technologies alter the production process, the inputs into the process or the product itself in such a way that it generates a smaller amount of, or a less noxious, residual. The Commission of the European Communities defines clean technologies as "any technical measures taken in various industries to reduce or even eliminate at source the production of any nuisance, pollution or waste, and to help to save raw materials, natural resources and energy".[5]

The question addressed in the following subsections concerns how governments can alter current technological development to stimulate the innovation of clean technologies. The quasi-evolutionary model described in the previous section suggests three general strategies to alter current lines of technological development (innovation as well as diffusion):

1   the development of alternative variations;
2   modification of the selection environment;
3   the creation or utilization of technological nexus.

The following sections elaborate these strategies, and discuss how governments use these strategies. The examples focus on Dutch experiences, but, where possible, these will be compared with those experiences of other European countries.

The development of alternative technologies

Governments could first of all try to develop, based on their own goals, alternative technologies which are not developed in the market. For example, in the field of energy, the Dutch government financed alternative technologies such as wind and solar power. However, these alternative technologies never survived commercialization, when market forces became dominant. Alternative technologies to guarantee the quality of labour suffered the same fate (Leydersdorff and Van den Besselaar, 1987).

The network approach provides an explanation for this. The selection environment is not receptive to these forms of technology, implying that for the successful commercialization of technology, the selection environment (infrastructure, measure of competitiveness, etc.) will also have to be changed. That is an arduous road to take, to say the least; not even the EDF's engineers were successful in this task.

It is quite clear that these changes cannot be brought about by funding alone. Another way to stimulate clean technology is to make R and D grants available. This is an attempt to steer existing innovation processes towards the desired direction. In the Netherlands, a clean technology programme was launched in 1975, in which industries and research institutes could apply for financial aid to develop clean technologies. Evaluation studies showed, however, that only a small part of the subsidy funds were actually spent on clean technologies. Furthermore, most of the subsidies went to larger firms which would have made the required investments even in the absence of such grants (Cramer *et al.*, 1990). Dutch experience, as well as that of other European countries, shows that the system of subsidies is far too weak an instrument by itself to be an incentive in the development of clean technology (Magat, 1989; OECD, 1985). A short comment must be made on this: the provision of grants can play a significant supplementary role in the deployment of other instruments. If, by means of strict standards, firms were forced to implement clean technology, grants could serve to minimize the economic risks and opposition to the severity of the standards.

Changing the selection environment

In setting and issuing regulations, the government influences a firm's selection environment. Governments make use of a wide range of instruments, the most important of which are imposing standards and levying charges. In general, the implementation of these instruments has led to the application of end-of-pipe technology (OECD, 1985; Cramer *et al.*, 1990).

There are two reasons for this. On the one hand, it is frequently easier for firms to meet the requirements without altering their production technologies. The standards and charges have never been strict enough to have had a technology-forcing effect, and regulations have mainly been aimed at resolving acute environmental problems, quick solutions to which were given preference. Thus the stimulation of clean technologies was primarily a derivative of a general environmental policy, and not of a specific technology-forcing policy.

In terms of the model presented earlier, we may assume that the policy instruments used have mainly had an *ex post* selection effect. In their technology development, firms rarely anticipate regulation, thus there is a lack of *ex ante* selection. Such *ex ante* selection is a prerequisite for the development of clean technology. To increase the *ex ante* effect, the setting of regulations should be oriented more towards the way in which variations are generated in firms. For example, factors such as "certainly in the long term" and the "time path" (phasing) are most important in technology development. This would mean that the government could set very strict standards, so stringent that the technology needed to comply with those standards is not yet available. However, these standards must be set and must remain unchanged for a long period of time, as well as being introduced in progressive stages. The government should also be flexible in its approach to implementation, sometimes making allowances for transgression over certain periods. Yet as far as the goal to be achieved is concerned, it must be unrelenting and unambiguous (Ashford *et al.*, 1985). Changing the selection environment through regulation will be particularly successful if regulations are designed in such a way that firms are able to anticipate their requirements, thus bringing about an *ex ante* selection pressure.

Further refinement and adaptation of the instruments is necessary, but not sufficient to stimulate the development and application of clean technology. Decisions as to whether or not to invest in clean technologies will depend on many factors and actors, of which government environmental regulation is just one. One of the biggest obstacles to the development of clean technology at present is that the government is more or less the only actor in the selection environment trying to impose environmental requirements. Government strategy should be aimed at encouraging other actors in the selection environment to set environmental requirements too. Such requirements should, in fact, form a regular element of market transactions between companies and, for example, banks and insurance companies.

I would like to give three examples of actors who are part of the selection environment and who could set environmental requirements for companies: insurance companies, manufacturers' associations, and companies contracting out work.

Insurance companies are increasingly abandoning their restrictive policy on insurance policies in respect of the environmental risks. In the Netherlands, a group of fifty-two insureres have set up a separate company for the purpose of jointly investigating the potential of this relatively new market. An increase is expected in the demand for insurance against environmental risks (Schot *et al.*, 1991). This is because there is an increased incidence of manufacturers being held liable for the risks arising from their products and processes. If the insurers start to explore this market in more depth, they are certain to impose various requirements, coupled with checks and inspections, on how companies behave in respect to pollution risks. Insurance companies could require reductions of certain risks, which could be achieved by clean technology. The government could speed up this process by tightening up liability legislation.

Manufacturers' associations are exerting more and more pressure on their members to comply with government regulations. Scandals can have an adverse effect on the image of a whole industry. Attempts are often made at industrial branch level to achieve a single line of policy. Manufacturers' associations, for instance the Chemical Manufacturers Association in the USA, are even starting to force their members to follow common policy under threat of expulsion. This works, because firms prefer to be recognized as acting responsibly (Schot *et al.*, 1991).

Another way in which firms exert pressure on one another is via their supplier-contracting firm contracts. This pressure can be even more intense, due to the present trend in Europe of contracting out more work to fewer suppliers (see Hagedoorn and Schot, 1988). Relationships with the remaining, smaller, number of suppliers are likely to be firmer, and those contracting work out are more able to impose their own quality control standards on their suppliers. So there is an increasing pressure on suppliers to deliver a certain standard of quality. Environmental demands could become part of that quality requirement, affecting mainly those medium-sized and small firms which the government can experience difficulty in reaching.

In short, the government could change the selection-environment in two ways to achieve a greater technology-forcing effect in the direction of clean technology. First, the *ex post* effect of existing environmental regulation must be changed to an *ex ante* effect. Second, the government could significantly broaden the selection environment; environmental requirements would then become part of regular market transactions.

Creation or utilization of the technological nexus

Stricter and broader selection pressures from the selection environment will not necessarily result in clean technologies. To achieve this, the requirements arising in the selection environment need to be linked with investment decisions taken in companies. This is a task for the technology nexus. The government should therefore attempt either to create these nexus, or to utilize existing ones. This is a strategy which up to now has not been exploited in a systematic way, either in the Netherlands or in the rest of Europe. I will discuss three examples of technological nexus which could be utilized far better by the government:[6]

1 marketing departments and other actors sensitive to public (or credibility) pressure;
2 environmental departments;
3 quality assurance departments.

Firms' marketing departments have the task of translating requirements from the selection environment into technological options. Marketing departments, therefore, are sensitive to indicators from the market, and also to public pressure. For example, Volvo translated diffuse public concern about safety into a specific design and sales strategy for safe cars. This forced other car manufacturers to either give more attention to aspects of safety, or risk losing credibility. Generally speaking, in the chemical industry credibility is increasingly experienced as a problem. This is apparent from the attention given to the environment in the annual reports of individual companies. The sensitivity of firms to publicity has also turned out to be a powerful weapon in the hands of environmentalists. It can be very damaging to a company to be accused of damaging the environment. The most well-known example of this in the Netherlands was Duphar. This company suffered enormous damage after containers contaminated with dioxin had been discovered, and had to launch a major publicity campaign to improve its image.

The environmental issue offers companies new opportunities in the field of marketing and image, while at the same time forming a threat to sales figures and that very same image. Governments could show a better understanding by encouraging improved articulation of the still quite diffuse pressure by environmentalists, the public and the consumer, and also by channelling this pressure in the direction of companies in question. The latter is quite possible by creating risk-communication obligations. At present this is the case in respect of a limited number of companies within the framework of the European Community post-Seveso directive.[7] American and Dutch experience indicates that the pressure to adopt cleaner production technologies is increased if a company starts to communicate with its immediate neighbours (Schot, 1990; Baram et al., 1990).

A second example of technological nexus is given by the environmental departments of larger companies. In the Dutch context, this is an example that shows how government policy can frustrate the genesis of a desirable nexus between variation and selection. Environmental protection is increasingly regarded as a separate function in companies. Institutionalization of the environmental care function takes place, and the outcome is that people are given the responsibility, in the form of a staff position or otherwise, for environmental protection within a company. Responsibilities become more clearly structured, and more is written down. A tentative professionalization of environment officers emerges. Slowly and cautiously, a shift is becoming apparent in the tasks of environmental departments. Three tasks can be outlined:

1 *Monitoring*: checking whether internal and external rules are observed, instruction and training is given, measurement and recording of emissions, waste products and processes, are conducted.
2 *External contacts*: contact with government bodies, the people living in the neighbourhood, environmentalists and the general public.
3 *Innovation and policy development*: influencing the company's strategic policy, for example in the construction of new factories or in determining R and D strategy.

At present, more amount of time is spent on the primary monitoring task and the subsequent contacts with government bodies. The environment officers of a company appear to contribute very little when it comes to the development of anti-pollution products, the generation of alternatives for scarce or pollutant raw materials, or the design of plants that would have less harmful effects on the environment. Thus, the third task plays a very minor role in the work of environment officers. The absence of an institutional link between environmental care and innovation is also apparent from the fact that in most cases the environmental aspect has no part in a company's strategic planning. Yet changes are under way in this area. Environmental departments are becoming more and more involved in decision-making on company investments. The Dutch government is at present taking very little advantage of this. Government policy on environment protection is focused on the enforcement of standards. Particularly because the government is presently unable to perform its enforcement task, more emphasis is put on self-regulation within industry. Environmental departments need to play a crucial role here. No attention has been given, however, to devising measures with regard to the position of environmental departments as the potential nexus between variation and selection. Nor are initiatives taken to strengthen the position of environmental departments in this context. On the contrary, because of the policy focus on enforcement of standards, the expectation is that environmental departments will be locked into their first task through the need to focus on compliance with government regulation. This means that the position of environmental departments with regard to investment decision-making and R and D strategy will not be developed.

The third and last example of nexus which can be influenced are quality assurance departments. In recent years, significant changes have taken place concerning quality assurance in companies (Hagedoorn and Schot, 1988). Quality has become a necessary and vital part of companies' marketing strategies. Companies which fail to comply with the required standards of quality are simply pushed out of the market. For many companies that contract out work, quality standards are a *condito sine qua non* for delivery. This has not always been the case; for many years, price was the only selection criterion in the supply market. Additionally, quality has started to take on a different meaning. In the 1960s and 1970s quality was monitored by quality control personnel or departments. These had the task of detecting, often by means of random checks, those products which failed to comply with the set standards of quality. Nowadays, quality is integrated throughout the organization. The method used to control quality has also undergone enormous change. It is no longer a question of random checks being made at the start of the process, half-way through the process and on the finished product. Monitoring now takes place as a continuous process of checking. In addition, safety often forms part of this quality assurance. This connection is made, for example, in procedures drawn up for various branches, such as good laboratory practice, good manufacturing practice, and quality assurance control. These procedures were drawn up in consultation between government and industry, and include commitments in respect of quality assurance systems. At present environmental requirements play only a marginal role in these procedures.

THE GOVERNMENT AS GAME REGULATOR

The aim of this chapter has been to make clear that an interdisciplinary synthesis of economic, sociological, and historical analyses of technical change could be used as a basis for managing that process. Neo-Schumpeterian and recent sociological and historical models were combined, leading to a quasi-evolutionary model of technical change. On the basis of this model, three basic strategies for influencing technical change were discerned and elaborated on: the development of alternative technologies, the changing of the selection environment, and the creation or utilization of technological nexus.

The three strategies complement each other. As argued, in addition to the necessity of larger and broader selection pressures to achieve substantial, further development of clean technology, it is also essential to couple those pressures to the technological development via the creation or utilization of technological nexus. Additionally, R and D grants should be deployed in order to help those firms overcome any financial obstacles. It can be argued that if governments elaborate these strategies, a fourth actor role will emerge alongside the three previously described roles of generating variations, selecting variations, and linking variation and selection. This fourth role is that of regulating the variation and selection process. Working out this regulator role can to some extent fall within the scope of what was said previously about the role of broadening the selection environment and the creation or utilization of nexus between variation and selection. However, some aspects in this role go a step further. The government should monitor the course of the technology game.

In a recent OECD document, this role was described as one of a "social creative regulator of technical change" (OECD, 1988, 22). In this form of regulation it is not so much a question of limiting an actor's freedom to take action. In this context regulation does not mean establishing or issuing regulations, but rather structuring action, making things work. The government could take on this role by creating suitable conditions for interaction and feedback between the various actors. The role of a creative "game" regulator can take shape through the creation of networks between actors; as a result, the interdependence between actors will increase. The pressure to utilize clean technologies would disseminate through the system more rapidly. The advantage would be that third parties could form driving forces for changing the "rules of the game", and thus the channels for influencing technology. In the longer term this could lead to environmental criteria becoming an essential part of the heuristics which steer technological change.

Notes

A version of this article will be published in *Science, Technology and Human Values*. For comments and suggestions I am indebted to Jacqueline Cramer, Dany Jacobs, Bas de Laat, Tom Misa and Arie Rip.

1   See the work of Staudenmaier (1985) for an overview of discussions within the field of history of technology. Discussion has only recently come into full swing in the sociology of technology. Overview works are: MacKenzie and Wajcman (1985) and Bijker *et al.* (1987). A similar discussion is also under way within Neo-Schumpeterian economics in terms of technology push versus market pull. For an overview of this debate, see Coombs *et al.* (1987), chapter 5.

2   Heuristics can tend to cluster around an exemplar. An exemplar may be seen as a subject which one tries to improve by applying a cluster of heuristics. The DC-3 is often cited as a perfect example. The advent of the DC-3 aircraft in the 1930s defined a particular technological regime; metal skin, low wing, piston engines. Innovation involved better exploitation of this regime (Nelson and Winter, 1977). The clustering of heuristics around an exemplar forms a paradigm. Paradigms will not, contrary to the assumption made by Neo-Schumpeterians, occur in all cases (see Van den Belt and Rip, 1987).

3   The social constructivist model (Pinch and Bijker, 1987) can be categorized as co-evolutionary as well. But their conceptualization of the seamless web between content and context differs from the network approach. In social constructivism, technology is formed through a process of interaction (variation and selection) between perceptions of relevant social groups. No technological development exists outside the perceptions of those groups. This approach eliminates the distinction between technological and contextual factors by conceptualizing the whole development of the technology as a *social* process. This was a reason for Law (1987b) to accuse social constructivists of social determinism, the counterpart of technological determinism. He emphasizes that the actor-network approach, unlike social constructivism, does not give preference to any factor at all. The development of technology is seen as a function of the interaction between heterogeneous elements; both of a technological and contextual nature. Technology itself can play a steering role, because it functions as an active, if non-human, actor. Bijker and Pinch (1987) defended themselves against this criticism by maintaining that the social element, contrary to what Law had suggested, has no special explanatory position. Technology may also play an explanatory role, but it is then a socially constructed technology — always seen through the eyes of a relevant social group. Notwithstanding Bijker and Pinch's counter-arguments, the social constructivism approach and the actor network approach clearly differ. In the actor-network approach, technology is developed through the interaction of elements in a network. Dynamism results from the interaction between those elements. During development, network-forming can come up against various obstacles, both technical and social. Technology can play an active role in overcoming these obstacles. The latter is impossible in social constructivism, for technology does not exist outside the perceptions of social groups.

Recently, Bijker has stressed the active role of technology as well. In his PhD thesis (1990) Bijker concludes: "The relations I have analysed in the previous chapters were both social and technical. Purely social relations are to be found only in the imaginations of sociologist or among baboons, and purely technical relations are to be found only in the wilder reaches of science fiction. The technical is socially constructed, and the social is technically constructed." Without recognizing it explicitly, Bijker is leaving here the Social Constructivist's frame.

4   This does not imply that that selection environment is unchangeable. On the contrary, as technology changes, the environment will adapt itself. The main point here is to argue that the selection environment has its own momentum, whereby it is able to exert influence on technological development.

5   See OECD (1987) for an overview of various definitions. There is no absolute contradistinction between clean technology and end-of-pipe technology, but rather a sliding scale from remedial to preventive measures. Generally speaking there is no either-or choice. The use of end-of-pipe technology will continue to be essential when utilizing clean technology.

6   These examples are partly based on a study on the "greening" of the chemical industry. In this study we investigated the role of environmental aspects in the strategy of actors both inside the company (R and D, marketing, purchase, quality control) and outside the company (accountants, banks, unions, environmental pressure groups). We made eight case studies of

multinational companies and interviewed seven small and medium-sized firms. Some results are discussed in Schot (1990). The full report is available in Dutch (Schot et al., 1991).

7 In 1976 in Seveso, Italy, a serious accident occurred at a chemical plant in which dioxin, a toxic substance, was released into the environment. In response to this accident, the Council of the European Communities issued on 24 June 1982, the so-called post-Seveso Directive. The Directive requires amongst other things that the public likely to be affected by an accident must be informed of safety measures, and the correct behaviour to adopt in the event of an accident. In the USA, similar developments can be discerned. In 1986 Congress enacted the Emergency Planning and Community Right to Know Act as Title III of the Superfund Amendments and Reauthorization Act. Title III calls for the establishment of local emergency response committees which, among other things, shall make provisions for public meetings to discuss emergency plans.

References

Akrich, M. (1988), "For an Anthropology of Technics", Paper prepared for the 4S/EASST Conference, Amsterdam, 16–19 November.

Ashford, N.A., Ayers, C. and Stone, R.F. (1985), Using Regulation to Change the Market for Innovation, Harvard Environmental Law Review 9, 419–66.

Baram, M.S., Dillon, P.S. and Ruffle, B. (1990), Managing chemical risks: corporate response to SARA Title III, Report of the Center for Environmental Management, Tufts University.

Belt, H. van den, and Rip, A. (1987), "The Nelson-Winter-Dosi model and the Synthetic Dye Chemistry" in The Social Construction of Technological Systems: New Directions in the Sociology and History of Technology (Eds W.E. Bijkwer, T.P. Hughes and T. Pinch) MIT Press, Cambridge, MA.

Bijker, W.E. (1990), "The Social Construction of Technology", PhD thesis, University of Enschede.

Bijker, W.E., Hughes, T.P. and Pinch, T. (eds) (1987), The Social Construction of Technological Systems: New Directions in the Sociology and History of Technology, MIT Press, Cambridge, MA.

Callon, M. (1986), "The Sociology of Actor-networks: The Case of the Electric Vehicle", in Mapping the Dynamics of Science and Technology (Eds M. Callon, J. Law and A. Rip) Macmillan, London.

Coombs, R., Saviotti, P. and Walsh, V. (1987), Economics and Technological Change, Macmillan, London.

Cramer, J., Schot, J.W., Van den Akker, F. and Maas Geesteranus, G. (1990), Stimulating Cleaner Technologies Through Economic Instruments. Possibilities and Constraints, UNEP Industry and Environment Review, 13(2) 46–53.

Dosi, G. (1988), Sources, Procedures, and Microeconomic Effects of Innovation, The Journal of Economic Literature, 26, 1120–71.

Dosi, G. et al. (eds) (1988), Technical Change and Economic Theory, Pinter Publishers, London/New York.

Hagedoorn, J. and Schot, J.W. (1988), Cooperation between companies and technological development, Centre for Technology and Policy Studies, Apeldoorn.

Hughes, T.P. (1983), *Networks of Power: Electrification in Western Society, 1880–1930*, John Hopkins University Press, Baltimore.

Lattour, B. (1987), *Science in Action*, Open University Press, Buckingham.

Law, J. (1987a), The Structure of Sociotechnical Engineering. A Review of the New Sociology of Technology, *Sociological Review*, **35**, 404–25.

Law, J. (1987b), "Technology and Heterogenous Engineering: The Case of Portuguese Expansion", in *The Social Construction of Technological Systems: New Directions in the Sociology and History of Technology* (Eds W.E. Bijker, T.P. Hughes and T. Pinch) MIT Press, Cambridge, MA.

Leydersdorff, L. and Besselaar, P. van den (1987), "What we have learned from the Amsterdam Science Shop", in *The Social Direction of the Public Sciences. Sociology of the Sciences Yearbook*. (Eds S. Blume *et al.*), **6**.

MacKenzie, D. and Wajcman, J. (eds) (1985), *The Social Shaping of Technology* Open University Press, Buckingham.

Magat, W.A. (1979), The Effects of Environmental Regulation on Innovation, *Law and Contemporary Problems*, **43**, 4–25.

OECD (1985), *Environmental Policy and Technical Change*, Paris.

OECD (1987), *The Promotion and Diffusion of Clean Technologies in Industry*, Paris.

OECD (1988), *New Technologies in the 1990s. A Socio-economic Strategy*, Paris.

Pinch, T.J. and Bijker, W.E. (1987), "The Social Construction of Facts and Artefacts: Or How the Sociology of Science and the Sociology of Technology Might Benefit each Other", in *The Social Construction of Technological Systems* (Eds W.E. Bijker, T.P. Hughes, and T. Pinch), MIT Press, Cambridge, MA.

Rip, A. and Belt, H. van den (1986), Constructive Technology Assessment: Influencing Technological Development? *Journal für Entwicklungs politik*, **3**, 24–40.

Rip, A. (1989), "Expectations and Strategic Niche Management in Technological Development (and a cognitive approach to technology policy)", Paper presented at the International Conference, "Inside the Black Box", Turin, 6–17 June.

Schot, J. (1990), "The Greening of Chemical Industry. An Assessment", Paper presented at the Center for Environmental Management, Tufts University, Medford, Massachusetts, 25 October.

Schot, J., Laat, B. de, Meijden, R. van der and Bosma, H. (1991), *Geven om de Omgeving. Milieugedrag van ondernemingen in de chemische industrie*, Line out Network, NOTA voorstudie nr. 15, Den Haag.

Schot, J. (in press), Constructive Technology Assessment and Technology Dynamics: Opportunities for the Control of Technology. The Case of Clean Technologies, *Science, Technology and Human Values*.

Rosenberg, N. (1982), *Inside the Black Box: Technology and Economics*, Cambridge University Press, Cambridge.

Slaa, P. (1988), *ISDN as a Design Problem. The Case of the Netherlands*, NOTA, The Hague.

Staudenmaier, J.M. (1985), *Technology's Storytellers: Reweaving the Human Fabric*, MIT Press, Cambridge, MA.

9

SHORT-TERMISM: CULTURE AND STRUCTURES AS FACTORS IN TECHNOLOGICAL INNOVATION[1]

Andrew Tylecote and Istemi Demirag

INTRODUCTION

This chapter deals with the causes, and technological effects, of short-term pressures in British manufacturing. We must first define "short-term pressures", and this in turn requires a definition of investment. An investment is any sacrifice of present (or earlier) cash flows, in return for improved future cash flows. The decision-taker's attitude to time clearly affects his willingness to invest. An economically rational organization will apply an infinite time horizon to all its investment decisions, discounting future revenues according to the cost of capital. Starting from economic rationality, then, short-term pressures could be defined as factors acting upon (or within) an organization which cause decision-takers (explicitly or implicitly):

1 to use a discount rate above its cost of capital; and/or
2 to choose some time horizon, beyond which future revenues are ignored altogether.

Short-term pressures will reduce the rate of investment below the economically rational level, and/or bias it towards short-term projects. But we may not be starting from a position of economic rationality, and so we do not know the initial discount rate or time horizon. A more robust definition of short-term pressures would be "factors tending to raise the discount rate applied (explicitly or implicitly) and/or to foreshorten the time horizon". Again, the effect would be to reduce the rate of investment and increase the bias towards projects with short pay-back periods. This latter definition allows inclusion of factors such as high interest rates and low profitability, which increase the opportunity cost of capital.

In this chapter we shall distinguish, first, between external and internal sources of short-term pressures, and examine their structural and cultural causes, and the different ways in which they may operate, taking two hypothetical examples. We shall see that short-term pressures acting within an organization are likely to be associated with pressures on middle and lower management to take a narrow, "sectional" view, avoiding efforts and expenditures whose benefits may go largely to other parts of the firm. We shall go on to show how they may work to affect the rate and pattern of technological progress. Finally, we shall use evidence from the early stages of an empirical investigation in which we are involved, to develop and refine a hypothesis on the causes and effects of short-term pressures in different UK industries.

DETERMINANTS OF SHORT-TERM PRESSURES

External pressures: a theoretical framework

A priori, one might expect the following factors to influence the extent of short-term pressures upon the firm:

1   *The cost of capital*: the opportunity cost of capital is determined by the availability, acceptability and period of external funds, as well as by interest rates; also by the extent to which external funds are required.

2   *The quality of information available to shareholders (and lenders)*: given perfect information, and a willingness to take account of it, one might expect shareholder pressure to be for economic rationality — the maximization of the present value of the firm's profits to eternity. To the extent that shareholders lack information relevant to the longer-term outlook, however — e.g. on technological progress "in the pipeline" — they will respond excessively to current profit and dividend figures, and similar data easily available.

3   *The objectives of shareholders*: these are likely to vary according to whether they are individuals, pension funds, banks, customers/ suppliers etc.; in particular the last two categories might well be interested in the growth of the company *per se*.

4   *The sensitivity of the firm to the share price*: shareholders' views of the firm's performance and prospects are reflected by the share price. Management will be sensitive to the share price to the extent that (a) it fears a hostile takeover, and (b) it wishes to issue new equity. It will be the more sensitive to (a), to the extent that it expects possible predators to be better informed than the market. Under (b) may come the wish to

carry out takeovers by an exchange of shares, as well as the raising of funds.

5   *The "tangibility" of desired investment*: investment can be categorized as tangible if it is on assets which can be capitalized, as "hardware" (plant and machinery) always can. To the extent that a firm's investment is tangible, there is no need for current profits to be reduced. What may be capitalized depends on the auditing regime. To the extent that it is permitted to capitalize research and development spending, the firm will be able to reduce the impact of this on current declared profits. (But note that this will only be relevant if we make very strong and adverse assumptions about factor (2): firms can always publish R and D expenditures, and leave readers to make their own allowances.) The capitalization of expenditures is no panacea — it immediately increases the capital base, which reduces the measured return on capital the following year; also, depreciation of capital assets will reduce profits. The protection from short-term pressures is itself only short-term. Much therefore depends on the next point.

6   *The predictability of the return on assets*: if at some point in the future it becomes clear that money spent has been wasted, it will have to be written off. Thus if R and D on such a project has been capitalized — as was the case in the 1970s with the Rolls-Royce RB211 jet engine — profits will lurch suddenly downwards. This will make the firm more vulnerable to a takeover bid than if the R and D had never been capitalized.

The incidence of external short-term pressures

Over time
In view of factors (1) and (4) above, it is clear that short-term pressures will be more severe during a cyclical downturn, particularly when that is accompanied by a tight monetary policy and high interest rates.

Across countries
Here it is mainly to factors (2), (3) and (4) that we should look. It is now an accepted distinction that German and Japanese shareholders act as (long-term) *investors* in shares, while their British and American counterparts tend rather to be (short-term) *traders* (see e.g. Charkham, 1989). This difference of approach, which is embedded in those countries' histories and social structures (Tylecote, 1982), directly affects the quality of information available to shareholders: German and Japanese shareholders, having in Albert Hirschman's terms forsworn "Exit", must choose the "Voice" strategy: they must find out how well

the firm is doing from a long-term perspective, and why, and if necessary do something about it — vote out the board, for instance. In order to economize on information and other costs, they will concentrate their holdings in a few firms. British and American shareholders, on the other hand, seeing themselves as traders, will diversify their portfolios so as to minimize risk and facilitate "Exit". Having a relatively small holding of each company's shares, they will find information costs relatively high. Moreover, for any individual trader, who seeks to make profit from arbitrage — trading based on knowledge that the price of an asset is different from its fundamental value — the value of information as to the long-term prospects of a firm (or its shares) is less than that relating to the short term. (For a proof see Schleifer and Vishny, 1990.) Factor (2) thus generates short-term pressures much more in Britain and the United States than in Germany and Japan. (On the over-reliance of the UK and US stock market on such information, see respectively Nickell and Wadhwani, 1987, and Joerding, 1988.) Moreover factor (3) varies between countries: key German and Japanese shareholders are often banks and customers/ suppliers, who may be expected to press for long-term growth as much as high profitability.

A detailed analysis would distinguish between the members of each pair. "The London market is perhaps the most short term of all" (senior fund manager quoted in IAB, 1990, p. 12). As to the pressures on the fund managers who manage the holdings of the largest category of UK shareholders, the pensions funds:

> the nominal performance measurement period has tended to shorten to two or three years, but in many cases the period is even shorter as trustees (or their consulting actuary advisers) focus on quarterly reviews of performances against the FT index or the median of all funds. The turnover of external managers appears to be high .. funds changed their managers every three years. (IAB, 1990, p. 13)

The IAB found gross dividend yields in the UK of 5.0 per cent, USA 3.5 per cent, Germany 1.9 per cent and Japan 0.6 per cent. In 1988, according to the same source, 64 per cent of all companies and 79 per cent of large companies believed the City did not take a sufficiently long-term perspective when making strategic evaluations of UK companies: in 1990 more than 90 per cent of UK finance directors considered the City to be excessively preoccupied with short-term earnings and share price performance.

The same considerations point to a similar difference in sensitivity to the share price. Charkham (1989) and IAB (1990) refer to the well-known vulnerability of UK and US firms, and the invulnerability of German and

Japanese, to hostile takeover based on a depressed share price. Likewise, though to a lesser extent, Anglo-Saxon firms depend more on equity finance, whose terms are more favourable the higher the share price.

Inter-industry differences

Some of the factors mentioned may be affected by the level of technology of the firm and industry. In a high-technology industry, that is one in which technology is sophisticated and quickly changing, the importance of intangible investment will be greater, and the predictability of return (on all investment) lower. To the extent that firms are growing faster, with more need for external finance, they will be more sensitive to the views of shareholders and lenders; the latter will also find it harder to make an accurate assessment of the firm's prospects. On the other hand, shareholders in a high-technology firm may recognize that its prospects depend heavily on its technological performance, and that they should not invest in it unless they are willing and able to assess this. Moreover high technology may give some protection from takeover: where the value of an organization lies mainly in the expertise of its employees — who may leave when they choose — it is harder to treat it as a commodity, to be bought and sold, and harder to fit it into the impersonal structures of a large diversified firm. Much, then, depends on the specific character of each industry.

Cultures, structures and the determinants of internal short-term pressures

In given circumstances, decisions at each organizational level are determined by the values of decision-makers (defined as a broad tendency to prefer certain states of affairs over others) (Hofstede, 1980, 19), and by their perceived circumstances. Thus if X desires to maximize profits (value) and believes policy A will lead to more profits than policy B (perceived circumstances) he will choose policy A.

Culture determines perceptions as well as preferences. Given "bounded rationality" (Simon, 1957) there is much scope to influence the pattern of search for information, and its evaluation once found. Thus (in the example above) Y's values may be identical to X's, but he may select and interpret the evidence available differently, concluding that it is B which is the more profitable option, which he therefore chooses. We thus follow Schein's definition:

> The . . . culture of any group or social unit is . . . the total of the collective or shared learning of that unit as it develops its capacity to survive in its external

environment and to manage its own internal affairs. Culture is ... taught to new members as the correct way to perceive, think about, and feel in relation to [internal and external] problems ... The power of culture is derived from the fact that it operates as a set of assumptions that are unconscious and taken for granted.

(Schein (1985), 19–20)

Like Gray (1988), we distinguish between culture at the level of whole societies, and subculture, at the level of an organization or profession.

By forming values and shaping perceptions, culture determines decisions in interaction with objective circumstances. Circumstances are partly determined by structures — understood broadly to mean structural forms and the rules under which they operate — which have the more effect on decisions, the lower the decision-maker's position in the organization. Those at the top of the hierarchy, who are less constrained, make decisions on structures (see Fig. 9.1), and have more scope to express their own culture.

We can show the interaction of culture and structure to produce short-term pressures in a hypothetical case, that of a large British company

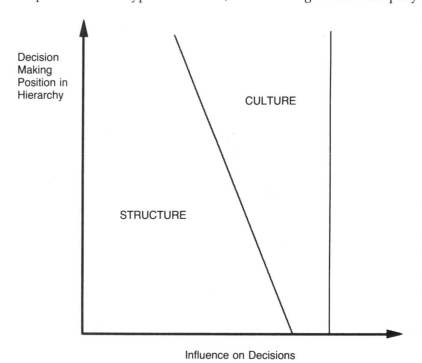

Influence on Decisions

Figure 9.1: Effects of cultural and structural factors on decision-making.

called Myopia Manufacturing plc. Let us begin with external pressures. Myopia has the typical pattern of shareholding for a major British company: its shares are mainly owned by an assortment of financial institutions, whose widely-diversified portfolios are mostly managed externally by professional fund managers based in the City of London and advised by investment analysts. This structural arrangement is presumably partly due to cultural factors: the City of London's lack of any tradition of expertise or intervention in domestic manufacturing industry (Tylecote, 1982). The same cultural factors interact with the structure to determine the *modus operandi* of the fund managers. The fragmentation, identity and behaviour of Myopia's shareholders are structural facts which its top management has to take largely as given — though it may be in their power over a period to "re-educate" the investment analysts and fund managers to judge Myopia by other criteria (e.g. new products in the pipeline), rather than by current profits, etc. If it has ever tried to do this, however, Myopia has been defeated either by the analysts' ignorance of technology, or their inability to persuade their fund managers that other fund managers will be influenced by such sophisticated notions. The result is that Myopia is under appreciable external short-term pressures.

How do Myopia's top managers perceive these pressures? They will surely be aware of their existence; but they may not realize how far longer-term performance will suffer if they yield to them. If for some reason they themselves evaluate the firm's performance in conventional financial accounting terms (Johnson and Kaplan, 1987), this may not occur to them at all. This approach might well be largely due to the system of remuneration (Drury, 1990). If top management's main stakes in Myopia are their salaries (and they expect to retire before long), profit-related bonuses, and stock options which they will soon be able to exercise, their personal self-interest may be little affected by the long-term performance of the firm. This distortion of goals may well induce a distortion of perceptions: they may not wish to know that their actions are not for the best in the longer term. Until recently, top managers' pay in Britain was much less tied to their firms' financial results or share price than in the United States (see Vancil, 1979 and Cosh and Hughes, 1987). However the trend has been to copy the US in this respect. Myopia has presumably done so; but why? If it is due to shareholder pressure, then we can say that at this level structure is shaping culture. But have we not argued that it harms shareholders' long-term interests? It may not, by comparison with the obvious alternative of pay simply according to position; and we have to distinguish between the long-term interests of (existing) shareholders, and the interests of long-term shareholders. The former may soon take their profits and run, like the managers.

It may be, on the other hand, that it is the top management who have taken the initiative; given the passivity of British shareholders it is probable. If top management are merely doing what they rationally calculate will please the shareholders, and thus strengthen the firm's position, the situation is little different from management by shareholder diktat. There may, however, be another explanation: that the background or environment of top mangement predispose them to believe that financial accounting measures are a good guide to the future. They may resemble the new breed of American manager, as denounced by Hayes and Abernathy (1980), taught on his MBA course to manage anything "by the numbers" without any interest in the technology. Or it may be a more specific case of organizational learning: TM may have taken over from a regime whose inattention to profits and cash flow had brought the firm to the brink of disaster. (GEC is an obvious case in point; see Jones and Marriott (1970), and Williams et al. (1983).) In such cases we can say that culture, or subculture, dominates structure.

If Myopia were a small firm, we could end the argument here: top management's time horizons would be the firm's. But Myopia is large and diversified: it has divisions, and these divisions themselves are divided into subsidiaries. The choice of strategies, and their implementation, will to some extent at least be determined by middle and lower management, whose goals and objectives will depend, again, on culture and structure. Structure is largely under top management's conscious control. The choice between M (multidivisional)-form and U (unitary)-form organization has considerable relevance to performance pressures. Myopia have chosen an M-form structure, of the extreme kind described by Goold and Campbell (1987) as the "financial control" style; it is not the only kind possible, in an M-form firm.

Figure 9.2 shows the organization chart for the financial control style:

The headquarters is slim, supported only by a strong finance function. Underneath, there are layers of general management, but prime profit responsibility is pushed right down to the lowest level. [Such] companies focus on annual profit targets. There are no long-term planning systems and no strategy documents. The centre limits its role to approving investment and budgets, and monitoring performance. Targets are expected to be stretching and once they are agreed they become part of a contract between the business unit and the centre. Failure to deliver the promised figures can lead to management changes.

(Goold and Campbell, 1987, pp. 46, 51–2)

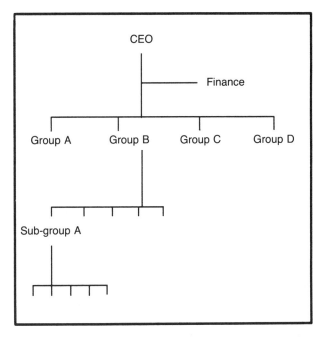

Figure 9.2: Manageable businesses' organization structure. (From Goold and Campbell, 1987.)

There are three possible reasons for Myopia's choice of this structure:

1 Because it is actually suitable to their business. ". . . This means that the businesses [which make up the financially-controlled firm] should have few linkages with each other, should be in relatively stable competitive environments, and should not involve large or long-term investment decisions" (op. cit., p. 46). The businesses which fit the bill are likely to be in declining, low-technology industries and thus — even if well managed — to have little prospect of organic growth. Goold and Campbell's sample showed well above average overall growth in 1981–5, but that was all due to acquisitions; there was organic shrinkage, though that may have been as much an effect of financial control as an indicator of fitness for it. Is any sector really low-tech, now? But in many it may for a long time be good for profitability (see below).

2 Myopia is under such acute external short-term pressure that its top management has truly no choice but to pass this on down the line to its subordinates: unless they can be induced to squeeze the maximum profit out of their activities, the firm will go bankrupt, be taken over, or be unable to find the resources for investment.

3   Myopia's top management falsely believes points (1) and (2) to obtain. It might believe (1) for the cultural or subcultural reasons referred to earlier.

We have described Myopia to show the possibilities for short-term pressures arising outside and inside British firms. For a contrast, we should describe another imaginary firm towards the other end of the spectrum, which we shall call Longsight plc. Longsight's management at every level is prepared to give full weight to prospects of profit a decade and more ahead. Clearly the pressures on and in Longsight are different from Myopia's. To take external pressures first, it may be that the divergence results from a different ownership structure. Thus in Oxford Instruments, a strikingly innovative company, a majority of shares is held by a small group of internal and external holders, all with a long-term commitment to the firm (Johnson and Scholes, 1989). But if that is a requirement, there will be very few (large) Longsights in Britain. Can we imagine a Longsight which manages to survive and prosper if owned by much the same assortment of City-based pension funds, insurance companies etc. as Myopia? It is possible that it benefits from being in a different industry, subject to more informed scrutiny; but credit for the higher quality of the scrutiny may be due to the efforts of Longsight and its competitors.

What adverse external short-term pressures it perceives, Longsight's top management resists, refusing — in the long-term interests of the company — to let its subordinates feel them. Accordingly, its relationship with them has nothing of the financial control style, conforming instead to Goold and Campbell's "strategic planning" style, in which top management involves itself closely in decisions on the development of new processes and products, monitoring their progress directly, providing a context in which a fall in a division's rate of return on capital can be understood and (where appropriate) accepted. In fact, divisionalization may not go so far in Longsight, which may centralize most or all of its R and D, at least. This protects R and D from any divisional temptation to economize on it, or to focus it on quick pay-back projects; however it does increase the difficulty of achieving good inter-functional coordination between R and D, marketing and production. If Longsight in fact achieves such coordination, this may reflect characteristics of the firm, or of its industry. The firm's corporate culture may have a certain interest in technology; its activities may not be very diverse; and it may have a practice of rather long-term employment. The industry may be one in which only limited inter-functional coordination is required.

EFFECTS OF SHORT-TERM PRESSURES: PROCESS AND PRODUCT INNOVATION

The effect of short-term pressures depends on what they are being compared with. If it is with economic rationality, then clearly investment and innovation in general will suffer. However, if the alternative is Leibenstein's x-inefficiency, or Cyert and March's organizational slack, there will certainly be some benefits, if the pressures are not too short-term. Where the management feel able to wait (say) two or three years to achieve visible improvements in performance, short-term pressure will provide a strong incentive to make changes which can yield results over that time-scale. It can easily be shown that the type of technological change likely to be favoured in such circumstances is process change: that is, improved methods of manufacturing an existing product, particularly where the improvement involves reductions in costs. The outlook for such process change is particularly favourable where worker and union resistance to required changes in manning and working practices is likely to be modest, and where the change required is not in the strict sense an innovation, merely the introduction of technology already introduced elsewhere. It can be shown, in the same way, that the type of technological change which will be most inhibited by short-term pressures is product innovation, particularly where the new product marks a radical departure from the existing range. Only exceptional organizational slack could be expected to lead to worse results in this area.

This latter conclusion is supported by an effect of the financial control style not yet mentioned: its tendency to generate pressures which narrow the spatial field of vision of decision takers. Those managing a given profit centre tend to concern themselves with whatever will improve their own results: cooperation with other parts of the firm will be given a lower priority. Technological progress which is facilitated by such cooperation will therefore be inhibited. This certainly affects product change, particularly radical product change: if the firm is to branch out in an altogether new direction, it will probably need to call on the expertise of a number of different divisions (Pavitt et al., (1989).

So far so good: if we know enough about the pressures acting upon and within firms, we should be able to make some predictions about the relative speed of the different types of technological progress. Now there is evidence that British industry has experienced during the 1980s a definite intensification of short-term pressures. This is certainly the consensus reported by IAB (1990) so far as external pressures are concerned, and this impression is supported by evidence cited there on the low rate of increase in industry-financed R and D during the period 1975–87, and the much higher rate of increase of dividend payments. High interest rates throughout the decade, and severe recession together with overvaluation of sterling at the beginning of it, were macro-economic factors making for short-term pressures. Privatization, deregulation, and the liberalization of procurement in the defence and telecommunications industry have presumably pushed in the same direction. Perhaps in response to external pressures, many large firms appear to have moved to, or towards, the financial control style. The pressure for quick improvements in profitability would therefore have increased; and the difficulties of making them via process change would have diminished, since:

1   Worker resistance appears to have been very weak, by comparison with the 1970s and even compared to (for example) France and Germany. (See Daniel (1987) and Batstone (1986). On the trade union legislation of the Conservative government, see Mayhew (1985).)
2   British industry was in general so far behind some of its competitors in productivity (Prais, 1981) that great advances could be made by technology transfer; and the growing interpenetration of Britain and its more advanced rivals by foreign and British multinationals has meant that much of this could take place within companies — the easiest way. (Much more process change was reported from foreign-owned than from UK-owned firms — Daniel (1987).)

One would therefore have expected some acceleration in process change, but some falling off in product innovation, to a rate much below those of our rivals less affected by short-term pressures.

The evidence available is consistent with these predictions. Muellbauer (1986) and others show a definite quickening of productivity growth in British manufacturing since 1980. The rate of product innovation, on the other hand, is not at all impressive: this can be inferred, for example, from Patel and Pavitt's (1988) Anglo-German comparisons of patenting rates for the late 1960s and the early 1980s, and (at the other end of the chain from R and D to output) to the alarming deterioration in the UK balance of trade on manufactures. (For the connection between patenting rates and subsequent trade performance, see for example Fagerberg (1988).) A recent survey of attitudes to R and D and the application of technology (PA Consulting Group, 1989) found that British firms put at least as much emphasis as their rivals in Japan, France, Germany and the Netherlands on investment in new technology as a source of new processes, but markedly less on it as a source of new products. Moreover, responsibility for R and D strategy tended to be at relatively low levels in British firms: in only 21 per cent did it lie with a main board director (57 per cent in Japan); in 22 per cent it was with middle management (Japan 1 per cent). This fits the picture of a relatively high proportion of "Myopias", with a segmentalized structure and financial control.

EARLY RESULTS

We have chosen to look at firms in three sectors: (ethical) pharmaceuticals, electronics and electrical engineering, and mechanical engineering, chosen as respectively:

1 A high-technology industry in which British-owned firms are highly successful in innovation, growth and profits. (See Patel and Pavitt (1987 and 1988), for evidence on innovation for the three industries.) Pharmaceuticals, and the chemicals industry with which it is associated, have been strikingly deviant during the late 1980s, increasing their R and D spending by 13.1 per cent per annum during 1985–8, while the average of other British manufacturing stagnated (IAB, 1990).
2 A high-technology industry in which British firms are much less successful in innovation and growth (and recently stagnant in R and D — IAB (1990)).
3 A medium-technology industry in which British firms are also rather unsuccessful, and in which R and D has also stagnated recently.

We have interviewed senior managers in two firms in pharmaceuticals and three across the engineering industries, which we discuss together.

Engineering

Our three firms are quite similar in history and present arrangements; we shall concentrate on one of them, E. E is a large, fairly diversified multinational whose activities range from high to decidedly low technology. Its principal product area involves it in relationships with customers in many countries and ranging from large nationalized industries for which its product is important, to small firms for which it may be a minor input. Like many other large British firms (Prais, 1981) during the 1970s its structure and control system were rather loosely defined. The "centre" was quite strong in numbers of employees and responsibilities, though much decision-taking, given its diversity, had to be delegated. In Goold and Campbell's terms, E was strategically planned in its major product area, with strategic control in the rest. In this form it was severely shaken by the deep recession at the beginning of the 1980s. E responded with far-reaching changes in top management personnel, who moved it to a well worked-out version of the financial control style. The centre's responsibilities were steadily pared down. The central labs were closed down and the technical director was moved out of the London HQ to that of the home division in the main product area.

The new system appears to be working very well, with rapid increases in profits and productivity over the five years to 1989, and good sales growth also. The productivity gains were to be expected from the move to financial control, which gives line management of product-based profit centres both the freedom and the incentive to deploy their resources to improve operating efficiencies; administrative and clerical jobs are cut, as the upper layers of administration are largely removed. Success currently protects E from external pressures. It is not afraid of takeover, and can self-finance all its investment and research and development. Since it covers low-tech as well as high-tech, R and D in the latter areas can be quite generously funded without seeming high overall, or substantially reducing profits. Well worked-out project proposals get a quick and usually favourable decision. Nor do they depend on divisional approval: funding can be sought from the centre.

We shall examine the new system's implications for innovation by way of two examples. The first is E's most successful innovation, now yielding large profits and promising more. After initial work in the late 1960s it committed itself in the early 1970s to major expenditure on R and D, in collaboration with a producer of the principal input, which had already made major advances. The strong interest and encouragement of the nationalized industry which was to be the main (domestic) purchaser of the output, completed a strong chain of user-supplier interaction, which Lundvall (1988) and others have found to be of great importance in innovation. The programme went well, and in the early 1980s the decision was made to move to volume production of the new product. The commitment of capital was large enough to gain scale economies and thus high margins, and the company established itself as one of the world market leaders in the area.

Disturbingly, some in the company believe that had the financial control system been in place during the 1970s, this project would not have been so successful. What is lacking in the new system of decision-taking is scope for project proposals without a well-argued set of estimates of costs and returns, but which may nonetheless be highly

profitable *if* they succeed. Nor can they be quietly begun without ___
approval; the research management structure is too lean for that:

> Under the new regime there is no central way of doing things ... previously we
> had a centre which had money and would spend it on central research, central
> planning, central market research. ... The centre can authorize an expenditure
> but there are no central people who can do this work ... so we have to go down
> the tree and say, "Frank, you are pretty good at such and such — will you find
> out a bit more about this [idea for a new product]". And the first thing his boss
> will say is, "Now look Frank — you are involved in this programme on [existing
> product]." ... the people down the tree are more risk-averse than they used to be
> because they are being driven harder for profit and cash. Although the boss at
> the top is saying "We are more willing to take risks", the people at the bottom are
> saying, "Well I know *you* may say that but I am being driven for cash and for
> profit."

Equally E's decentralization, however necessary, tends to cause neglect
of projects which need to draw upon the expertise of a number of
divisions.

Our second example may illustrate the effect of the new system —
with the caveat that it was chosen as an example of failure. In the main
product area there is a fairly new technique, widely applicable, in
principle, to improve certain product characteristics; but if it is to be
used, there must be in each case a degree of innovation in process and
product. It can only be used in a purpose-built facility with high capital
costs — though low variable costs and high capacity.

Recognizing the scope this technique offered to move many of E's
products up-market, the technical director persuaded the centre in the
early 1980s to buy a facility which would potentially be widely useful in
different divisions of the firm. The facility was sited near one of the main
domestic manufacturing sites in the main product area. As the control
system was then being implemented, it was allocated, for accounting
purposes, to the corresponding division — and then, as the system was
further decentralized, to one of its subdivisions. The management of
this subdivision had then to sell the facility's services to other profit
centres. They did not price them at marginal cost, since the fixed costs
were so high, but at average standard cost, assuming normal capacity
utilization.

This defect in the shadow pricing system was compounded by a
defect in the response of potential users within the company. These
were mainly in subdivisions producing a wide range of products in
smallish volumes for numerous industrial customers. Accordingly, the
user-supplier interaction took place at a relatively low level. Those who
should have been discussing with their customers how they might move

up-market with the new technique, had the primary task of selling them the existing products, and it seems did little else. Why?

> [In this part of our business] you are supplying into ... little individual niches, they are not controlled by the larger market. [The buyers] are all saying, "That is what I want — give it to me ... I want it today". ... I think our marketing culture was geared to that business. What we didn't have was the foresight somehow to have a duality of marketing approaches, i.e. we will give you what you want today but let's keep talking about tomorrow .... The business has never been particularly profitable and you couldn't afford to spend some time thinking about tomorrow because the boss is saying, "Look, we have got this product, let's sell it."

As another informant explained, some rivals did better:

> The best of the competition tend to say: "Here is a product that is [superior in certain properties] — let us find someone who needs these properties and persuade them ...". When that product emerges we say, "Ah, we have got to copy it", because the market now wants [a product with these properties].

Why had E failed to do this? A cultural explanation might be that those involved were not inclined by training or tradition to take on an extra innovative role. A structural explanation would stress our informants' reference to the short-term profit pressures on the line and marketing managers concerned. The system of rewards and evaluation would make the manager of the profit centre regard the gains from an innovation as remote; to his marketing managers they would have been virtually invisible.

The contrast between our two examples reflects not only a change over time, but also a difference in the level at which user-supplier interactions were bound to take place. Those on which the successful innovation depended were from the beginning at relatively high levels in the firm, because of the obvious importance of the matter, and because of their relationship to both user and supplier. Even now, the performance pressures at higher levels will be less acute. Likewise, with the "failure", the low level at which user-supplier interactions were required would have caused problems in the 1970s, though perhaps for rather different reasons. The system then sent weaker signals, so that there was less pressure in any direction.

Pharmaceuticals

The ownership pattern of our two firms appears to be quite typical of British industry. In spite of their companies' efforts to "educate" the investment analysts advising their owners on them, our informants did not feel much had yet been achieved: the analysts had only a most limited ability to judge the value of products "in the pipeline":

> I think their valuations of the business and profits are in fact long term but …
> inappropriately influenced by the more recent single events. … A positive or
> adverse finding in a single clinical study will raise or lower the share price and put
> pressure on profit expectations. … The problem is thus a market which is over-
> sensitive and reacts excessively to every result, rather than a market which
> operates in a more damped fashion with a view as to whether the long-term
> strategic target is still achievable within the forecasted timetable.

The companies could gain nothing even in the short term by reducing R and D expenditure. Our informants believed it would have led to a fall in the share price, because analysts had come to expect a certain, high, R and D intensity from them and accepted that it had something to do with their high profits. (A substantial increase would also have upset the market, however!) No doubt if such expenditure had required a cut in dividend, the reaction would have been different: both were obviously profitable enough to afford it.

The analysts may have learnt over the years from the experience of the industry: in ethical pharmaceuticals, firms live or die by product innovation, on an international market where there is little or no protection through tariffs or otherwise. They must risk their own money on these innovations, and may lose every penny up to and even beyond the full introduction of a drug on to the market. Because of the duration of clinical trials, the period from initial spending in research, into a possible drug, to a positive cash flow, is unusually long. Only those firms which have spent heavily (and wisely) on innovation, have prospered. Some understanding of this in the City can be inferred from a study quoted by IAB (1990, p. 7) which found that sixteen out of eighteen pharmaceuticals analysts were science graduates; that was true of only eight out of twenty-five electronics analysts, despite that industry's research intensity.

Performance pressures within pharmaceuticals firms are likely to be affected by another unusual feature: the *manner* of innovation. Successful innovation normally requires close links among at least three functions within the firm: R and D, marketing and production, from an early stage right to the end. Close coordination is certainly required in engineering; British culture tends to prevent it. The tension between these facts may explain the extreme decentralization of many British engineering firms. Pharmaceuticals seems to be at the opposite extreme: R and D can operate for long periods almost alone. The main interaction required is with marketing, but even this is limited:

> In the pharmaceutical industry it is not possible to modify the product during
> development. A choice is made when going from R to D, and then the properties
> of the compound are in fact fixed by the activities of the drug. Marketing can

choose to emphasize a particular activity but cannot build it into the drug if it does not exist. It is for this reason that marketing/R and D interaction is strong at inception of R and D and at compund selection, but that during development, marketing must play a limited part. (I think this is in contradiction to engineering where the product can be modified almost continuously, or at least to a very late stage.)

The interaction of R and D with production is minimal: once the researchers have synthesized (or discovered) a substance which appears to "work", they ask the production department to provide the successively larger volumes required. Nothing that that department tells, or could have told them has much implication for their work.

Thus inter-functional coordination is made almost unnecessary by the character of the industry. This had made it possible for firms to retain a U-form structure — divided primarily by function — rather than adopt the M-form, where the primary division (sometimes even the secondary) is by product. Clearly the coordination of functions is easier in relatively small product divisions, rather than across the whole firm. On the other hand, inter-*divisional* coordination between products then becomes the greater problem; such coordination can take place within functional departments, in the U-form firm. As we have seen, product divisions can easily transmit, or even generate, short-term financial performance pressures, and such pressures will further inhibit inter-divisional coordination. Such problems do not arise in the U-form pharmaceutical firm. Its innovative core, its R and D, is not a profit centre but a *cost* centre, and it is responsible to top management.

We have found an industry, then, which has little to lose from one old British vice — poor inter-functional coordination — and can thus protect itself from two new ones — poor inter-divisional coordination and internally-generated financial pressures. Even the City's notorious myopia has little power to hurt it. The problem with the City, after all, is not that it does not *care* about long-term profits, but that it lacks the expertise to predict the effect on them of current policies, thus providing an incentive to adopt policies which sacrifice long-term to short-term profits. Now the only such policy in pharmaceuticals would be a cut in R and D, and even the City has the expertise to predict the effect of that.

UK pharmaceuticals seem to be fortunate in two other respects:

The user-supplier relationship
UK firms' main domestic users are of course in the National Health Service, and among them they have been able to find collaborators, in the major teaching hospitals, who were themselves not under short-term profit pressures, but able to formulate and advance towards

long-term goals, to a greater extent than might be expected under a private healthcare regime. (Whether this situation can continue in the face of the recent decline, by international standards, in NHS funding, or the planned moves towards market disciplines, is an open question.)

British culture

The British educational tradition elevates pure science over applied science, and stresses the goal of knowledge rather than efficient production. One result has been more and better physicists and chemists, fewer engineers. This must hurt the engineering industries; but for fine chemicals and pharmaceuticals, the key sources of innovative manpower are prestigious pure sciences, chemistry and biochemistry — allied with the equally prestigious medicine. Moreover, the desire for knowledge for its own sake is less adverse to successful innovation in pharmaceuticals than elsewhere: there is much less need for people in R and D to keep responding to the requirements of the customer and the production engineer in collaboration with other departments.

Pharmaceuticals may be quite well suited to another, newer feature of British culture. International comparative studies have been made recently (NEDC Long-Term Perspectives Group, 1987) of labour force preferences among three principal goals: (1) material affluence and security; (2) the respect and acceptance of other members of one's organization or group; (3) self-fulfilment. Among the countries studied, only the Netherlands scored higher than Britain in emphasis on self-fulfilment. The pharmeceuticals researcher, whose work can offer only a long-term and uncertain pay-off, may need such a bias.

CONCLUSIONS

The incidence of short-term pressures

External short-term pressures

Clearly UK industry, like that of the USA, can be expected to experience more short-term pressures than that of Germany and Japan. Among sectors and firms we would expect considerable variation. A strong liquidity and profitability position would give some protection; it would also help if there were few or no "predators" in a good position to value and bid for the firm. If the firm's value largely depended on the skills of a few key people, that would also provide some protection against predators. Short-term pressures would bear less hard on technological progress where — as in pharmaceuticals — the market had learnt its worth in profits and could see also that it was very closely associated with high research and development expenditure. Clearly recession and/or tight money would exacerbate external short-term pressures.

Internal short-term pressures

The main suspect here is the M-form firm, and in particular its more extreme variant, the "financial control" style. This appears to generate pressures for foreshortened time horizons, and narrowed, sectional objectives, at middle and lower levels of management.

The user-supplier relationship

This has been shown to be highly important to successful innovation, and can usefully be brought within the analysis of the effects of performance pressures. It is affected by structural factors — the level and type of management which is engaged in the relationship and the pressures upon them — and cultural ones, which include the existence (or otherwise) of a common language of (dynamic) technology among those who conduct the relationship.

Reflections

These conclusions suggest some reflections on the economist's approach to such matters. Mainstream economics has only recently begun to look inside the "black box" of the firm, with the rise of "new institutional economics". It has done so carrying some of its usual baggage, individualism and functionalism. The first has led it to translate the collectivity of the firm into a web of individual contractual relationships, which leads, if a web can be said to do so, upwards to the board of directors and thence to the shareholders. The parties to these contracts are of course motivated by self-interest. The second involves the assumption that what is — or at least, the direction in which we are heading — is best, for, to quote Jensen and Smith (1985):

> Since most goods and services can be produced by any form of organization, different forms of organization compete for survival in any activity just as different species compete for survival in nature. . . The form of organization that survives in an activity is the one that can deliver the product demanded by consumers at the lowest price while covering costs.
>
> (p. 97)

They would no doubt agree to stretch the Darwinian analogy to allow for organizational learning and for the market in corporate control, which together can allow a successful species to gain ground much more quickly than under natural reproduction. The functional result would come all the quicker. The same confidence is expressed in the workings of the market: markets are too "efficient" to exert external short-term pressures. Yet in fact one of the main tools of their analysis can be used to help reach quite different conclusions. Following Williamson (e.g. 1981), new institutional economics generally lays stress on the problems of incomplete information within contractual relationships. Within it, agency theory focuses on the specific role of contractual provisions in modifying the behaviour of agents in a principal/agent relationship, such as that (arguably) between shareholders and top management (Thompson, 1988). Short-termism in Britain can be seen as expressing an acute case — or rather a chain of cases — of the principal/agent problem. If we begin with the pensions funds, the trustees, as principals, have great difficulty in evaluating the performance of their agents, the fund managers.[2] They choose a measure which will give them the illusion of control, such as three-monthly changes in fund values. The fund managers in turn play principal *vis-á-vis* their agents, the top managers, and again often have no better information on which to judge their performance than a few short-term numbers. Matters may be little better, down through the various tiers of management of the M-form firm.

The response of the agency theorist is that the contracts at various levels have to be carefully designed to maximize the agent's incentive to optimal behaviour from the point of view of the principal. Thus at the top of the firm, profit-related bonuses and stock options will tend to align managers' objectives with those of shareholders. The M-form firm then provides the vehicle for top management, suitably motivated, to control the rest of management in the shareholders' interests. It therefore comes as no surprise that this organizational form has been spreading rapidly across the world from its home in the USA, and that on various measures of profitability it has thoroughly outperformed its competitors (Cable, 1988). But Cable's survey carries a small reservation; like Mercutio's wound, not so wide as a church door, nor yet so deep as a grave, but 'tis enough, 'twill serve: the German and Japanese results do not fit. In Germany and Japan, M-form firms appear on the average to have lower profits than their rivals. Given these countries' record of success in international competition, we have to ask whether there is another species — one or more — in these countries which will ultimately supplant the M-form, as the mammals did the dinosaurs. And if so, why is it not doing so now, in the English-speaking world?

At this point we must look to sociology and organization theory rather than economics. The contingency theory of organizational structure denies that there are any universally valid rules of organization and management (Burrell and Morgan, 1979). Natural models of organization treat them as phenomena which display rationality and purpose only to a limited extent; their behaviour can be better explained in terms of power and processes of interaction (Otley, 1988). (We might also add culturally-determined prejudice.) Is not Jensen and Smith's Darwinianism a sufficient riposte to the natural models — the fittest will survive, even if few are trying to become fit? It depends, *inter alia*, on the accuracy of the selection process. If that is being operated by fund managers, rather than God, it may not be great. Who will ever know whether Plessey (strategically planned) was better run than GEC (financially controlled)? — the fund managers sold their Plessey shares to GEC. GEC adopted financial control in the 1960s, in what may from a long-term perspective have been a case of mistaken organizational learning.

One key difference between the English-speaking countries and Germany and Japan may be in culture. Culture strongly influences the behaviour of financial institutions and the stock market, and thus the performance pressures which they impose upon firms. It also influences the way managers generate, and respond to, performance pressures at various levels in the firm. The appropriate culture may lead top managers to define objectives of long-term organic growth for the firm which imply a strong commitment to product innovation (it will of course help them to defend such objectives if they can argue convincingly that they are in the shareholders' interests!). But just as culture affects the behaviour of top managers, so their behaviour and the structures they create influence the culture, or the subcultures, of their subordinates (as we saw with E's marketing managers). One aspect of culture is of course language: if there is a language of technology, and technical progress, which is understood and accepted throughout the firm, then it will offset the tendency of the language of accounting to generate short-term pressures. (At the opposite extreme, only R and D and production have such a language, and not even the same one.) In such ways German and Japanese firms, and their shareholders, may avoid most of the agency problems which arise from a combination of conflicting goals and inadequate information.

The economist may reply that there are perfectly straightforward structural reasons for the difference in behaviour of German and Japanese firms: ultimate power over them, as pointed out earlier, is often held by banks, customers and suppliers, whose objectives as institutions would in any country be more oriented to long-term growth than to profitability, short-term or otherwise. Perhaps; but why? Why have these institutions built up share stakes — in many cases quite recently? Why in Japan do the other institutions with strong shareholdings — investment trusts etc. — think it proper to behave as long-term but passive investors? Why have German families like the Siemens chosen to maintain strong share stakes and corresponding influence in major German firms over many decades? Culture dominates.

Notes

1 The research described here is part of a project on "Performance Pressures and Technological Progress in British Industry", funded by the SERC/ESRC joint initiative on "The Successful Management of Technological Change", whose support, together with that of the Sheffield University Research Fund, is gratefully acknowledged. We are also obliged to Brian Twiss and David Budworth, past and present "facilitators" of the SERC/ESRC committee, for their help and advice; and to the editors of this book, and Brian Loasby who was discussant at the seminar, for their helpful criticisms of the original paper. (Our team at Sheffield University is led by Tylecote and co-coordinated by Demirag, both of the Management School; it also includes Professor I.L. Freeston (Electronic and Electrical Engineering), Professor R.A. Smith (Mechanical Engineering), Dr N.D.S. Bax (Pharmacology), Dr Michael Kirk-Smith and Mr Ben Morris (Management), who have contributed much to this study.)

2 We pass over the highly unsatisfactory initial relationship between contributors and trustees, which introduces further complications, and arguably another link in the chain.

References

Batstone, E.V. (1986), Labour and productivity, *Oxford Review of Economic Policy*, 2(3), 32–43.

Burrell, G. and Morgan, G. (1979) *Sociological Paradigms and Organisational Analysis*, Heinemann, London.

Cable, J. (1988), "Organizational Form and Economic Performance", in Thompson and Wright (eds).

Charkham, J. (1989), "Corporate governance and the market for companies: aspects of the shareholders' role", Bank of England Discussion Paper no. 25, March.

Cosh, A.D. and Hughes, A. (1987), The anatomy of corporate control: directors, shareholders and executive remuneration in giant US and UK corporations, *Cambridge Journal of Economics*, 11, 401–22.

Daniel, W.W. (1987), *Workplace Industrial Relationships and Technical Change*, PSI/ Frances Pinter, London.

Drury, C. (1990), Lost Relevance: a Note on the Contribution of Management Accounting Education, *British Accounting Review*, 22, June, 123–35.

Fagerberg, J. (1988), International Competitiveness, *Economic Journal*, **98**(391), December, 355–75.

Goold, M. and Campbell, A. (1987), Managing Diversity: Strategy and Control in Diversified British Companies, *Long Range Planning*, **20**(5) 42–52.

Gray, S. J. (1988), Towards a Theory of Cultural Influence on the Development of Accounting Systems Internationally, *Abacus*, **24**(1), 1–15.

Hayes, R.H. and Abernathy, W. (1980), Managing our way to economic decline, *Harvard Business Review*, **60**(3), 70–80.

Hofstede, G. (1980), *Culture's Consequences*, Sage, Beverly Hills.

IAB (Innovation Advisory Board) (1990), *Innovation: City Attitudes and Practices*, Department of Trade and Industry, London.

Jensen, M. and Smith Jr C.W. (1985), "Stockholder, manager and creditor interests: applications of agency theory", in *Recent Advances in Corporate Finance* (Eds E.I. Altman and M.G. Subrahmanyam), Richard Irwin.

Joerding, W. (1988), Are stock prices excessively sensitive to current information?, *Journal of Economic Behaviour and Organization*, **9**(1) January, 71–87.

Johnson, G. and Scholes, K. (1989), *Exploring Corporate Strategy: Text and Cases*, Prentice-Hall, London.

Johnson, H.T. and Kaplan, R.S. (1987), *Relevance Lost: The Rise and Fall of Management Accounting*, Harvard Business School.

Jones, R. and Marriott, O. (1970), *Anatomy of a Merger: A History of GEC, AEI and English Electric*, Cape, London.

Lundvall, B-A. (1988), "Innovation as an Interactive Process: User-Producer Relations", in *Technical Change and Economic Theory* (Eds G. Dosi *et al.*), Pinter, London.

Mayhew, K. (1985), Reforming the Labour Market, *Oxford Review of Economic Policy*, **1**, Summer, 60–79.

Muellbauer, J. (1986), The Assessment: Productivity and Competitiveness in British Manufacturing, *Oxford Review of Economic Policy*, **2**(3), i–xxv.

National Economic Development Council Long-Term Perspectives Group (1987), *IT Futures — It can work*. NEDO, London.

Nickell, S. and Wadhwani, S. (1987), Myopia, the "Dividend Puzzle" and Share Prices, *Centre for Labour Economics, LSE, DP* no. 272, February.

Otley, D. (1988), "The Contingency Theory of Organizational Control", in Thompson and Wright (eds).

Patel, P. and Pavitt, K. (1987), The Elements of British Technological Competitiveness, *National Institute Economic Review*, **122**.

Patel, P. and Pavitt, K. (1988), "Technological Activities in FR Germany and the UK: Differences and Determinants", SPRU Working Paper, March.

Pavitt, K., Robson, M. and Townsend, J. (1989), Technological Diversification and Organization in UK Companies, 1945–83, *Management Science* **35**(1), 81–99.

Prais, S.J. (1981), *Productivity and Industrial Structure: a Statistical Study of Manufacturing Industry in Britain, Germany and the United States*, Cambridge University Press, Cambridge.

Schein, E.H. (1985), "How culture forms, develops and changes", in *Gaining Control of the Corporate Culture* (Eds R. Kilmann *et al.*), Jossey-Bass, San Francisco and London.

Schleifer, A. and Vishny, R. (1990), Equilibrium Short Horizons of Investors and Firms, *American Economic Association Papers and Proceedings*, **80**(2) May, 148–53.

Simon, H.A. (1957), *Administrative Behavior*, Macmillan, New York.

Thompson, S. (1988), "Agency Costs of Internal Organization", in Thompson and Wright (eds).

Thompson, S. and Wright M. (eds) (1988), *Internal Organisation, Efficiency and Profit*, Philip Allan, Oxford.

Tylecote, A.B. (1982) "German Ascent and British Decline, 1870–1980", in *Ascent and Decline in the World-System* (Ed. E. Friedman), Sage, Beverly Bills.

Tylecote, A.B. (1987), Time Horizons of Management Decisions: Causes and Effects, *Journal of Economic Studies*, **14**(4), 51–64.

Vancil, R.F. (1979), *Decentralisation: Managerial Ambiguity by Design*, Dow Jones-Irwin, Illinois.

Williams, K., Williams, J. and Thomas, D. (1983), *Why are the British Bad at Manufacturing?* Routledge & Kegan Paul, London.

Williamson, O. (1981) The modern corporation: origins, evolution and attributes, *Journal of Economic Literature*, **19**, 1537–68.

10

INDUSTRIAL TECHNOLOGICAL INNOVATION: INTERRELATIONSHIPS BETWEEN TECHNOLOGICAL, ECONOMIC AND SOCIOLOGICAL ANALYSES

Philip J. Vergragt, Peter Groenewegen, and Karel F. Mulder

INTRODUCTION

In this chapter we explore ways of integrating sociological and economic approaches to the analysis of technological innovation in industrial firms. Until recently, analyses of industrial innovation were made in either economic or social terms. However, we argue that both economic and social elements have a role to play in the analysis.

In microeconomic and managerial analyses, the focus is on the organizational structures of firms determining the success or failure of technological innovations. In evolutionary economics, the emphasis is rather on the interaction between the technology and its external environment. In social constructivist theories of technological innovation, the external social actors seem to play a dominant role in the shaping of technologies, but the economic aspects are not included in the analysis.

We shall focus on the interrelationships between technological innovation, organizational structure, and the external environment of the firm. Research on the relation between organizational structure and the innovation process usually examines the influence, in economic terms, of organizational characteristics on technical success. Much less attention is paid to technological development as a process in which internal and external rearrangements of the organization occur.

A specific interest of this chapter is the development of large-scale technology. We argue that this type of innovation consists of the parallel development of technology, organization, and the environment. Technological projects interact with the strategic outlook and existing capabilities of the organization. Changes in the technological project have to be accompanied by changes in inter- and intra-organizational relations.

As an example, we analyse the technological innovation of certain high performance fibres by several large industrial corporations, examining three dimensions and their interrelationships: the technological development itself (the technical characteristics of the innovation); the internal structures of the firms; and the external relations (both market and non-market) of the firms. The case of the development of strong fibres is used as an example to show how technological innovations, when they are created by chemical firms, are influenced and modified by social processes inside and external to the firm, and to indicate the role of economic considerations in the choice-making processes. We shall analyse the economic and non-economic considerations used in corporate decisions, concerning both the technology and the organizational rearrangements.

In this analysis we demonstrate that, in the case of large-scale industrial technological innovation, the technical characteristics of the innovation are closely interrelated with the economic aspects as analysed by managers, but that they are similarly influenced by the inter- and intra-organizational arrangements and rearrangements.

The description of the firm as an actor seemingly has a fit with micro-sociological approaches to the innovation process (see Bijker *et al.* (1987), Vergragt (1988), Law and Callon (1988)). In this theoretical approach, choices are the result of social processes inside and around the firm, in which social actors (organizations, organizational units, groups or persons) share expectations about the technology, form alliances in order to promote the technology, and try to enrol other actors in the technology. However, there is a problem with integrating this type of approach in a theoretical framework derived from managerial and economic studies.

In order to proceed in this way, we elaborate the existing theoretical traditions in the studies of technological innovations. We evaluate the managerial, social-constructivist, and evolutionary economic theories for their utility in explaining choice-making in industrial innovation.

We go on to argue that a characterization of the innovation process can be achieved, distinguishing three dimensions. Company strategy consists of a specific combination of actions on these dimensions, which are technological development, organizational structure, and the corporate environment.

This conceptual scheme is applied in a study of new materials innovation in large chemical companies. We use five case studies as examples to illustrate the dimensions in specific situations. The cases are chosen in such a way that there are differences in some of the dimensions, and similarities in others. By this approach, we hope to show that each of the three dimensions is of interest in technological choice-making.

We discuss at the end of the chapter the possibility of integrating economic and social theories on technological innovation processes. We will suggest some modifications, in order to take into account the interaction processes between technological innovations, organizational structures, and the external environment.

THEORIES OF TECHNOLOGICAL DEVELOPMENT

Traditionally, technological innovation has not been included in economic theories. Technological development was regarded as an independent, exogenous factor, accounting in an unspecified way for some element of economic development.

In evolutionary economics (see, for instance, Coombs et al., 1987), technological innovation has become a central theme that is used to explain the disequilibrium that forms the basis of economic change. Schumpeter analysed the innovative behaviour of entrepreneurs, exploring the way in which technical ideas for new products led to "gales of creative destruction". Companies compete by making new technological products and processes, rather than on the basis of price alone. This technological innovation can lead to a (temporary) monopoly, which can then be exploited. An extension of this theory has led to a set of loosely connected ideas about the mechanisms responsible for change. Companies develop research strategies that create a certain degree of order in their development of technological advantages. These actions may be structured, resulting for a specific company in a technological path, which may be further explored as a result of various factors. During their development, technologies improve and market relations are established, resulting in "learning by doing", by which technologies become more efficient and better adapted to the market. It has been reasoned that engineers have certain conceptual frameworks, or technological paradigms, for the solution of technical problems. As a result, a certain degree of path-dependency of a technology develops: once moving in a particular direction, a technology tends to continue to develop in the same direction. At the macro-level, this amounts to Nelson's "natural trajectories", and "technological" regimes such as miniaturization, mechanization, or automation (Nelson and Winter, 1982). At the micro level of the technology or the technological system, this means that alternatives are not lightly considered. However, within this paradigm new variations are tried out. The selection environment, which includes both the market and government regulations, chooses the most appropriate technology.

In management and organization theory, the unit of analysis is the innovative firm. In management theory, there is a concern with the distribution of organizational resources that are relevant to technological innovation. Examples can be found in the discussion of such issues as the amount of money to allocate to R and D, the organizational structure in which to integrate innovation management, the career possibilities for scientists and engineers, the communication pattern, the style of management, and the economic criteria by which decisions can be supported (Twiss, 1980). Organization theory increasingly addresses the role of the environment in the survival of firms. Abstract notions such as markets are increasingly depicted as social relations between firms (Granovetter, 1985; White, 1981). This has led to a lively debate in the overlapping area between economics and organizational theory, focusing on the role of hierarchical and market relations between organizations as mechanisms for the survival of firms (Williamson, 1986; Thorelli, 1986).

In social-constructivist and recent historical theories (see for instance, Callon (1986) and Pinch and Bijker (1984)) of technological development, the emphasis is on the actor perspective. An actor is either a person or a social group, which performs some type of activity or action. The reason for these actions is that actors have aims and interests. Pinch and Bijker argue that actors give some "meaning" to a technology, which requires that the technology performs a certain function. If the function is not adequately fulfilled, the actor tries to influence the technological development or artefact in such a way that these needs are fulfilled. This is seen as the basic mechanism by which technology develops. Callon explains that actors have to define problems, which are problematizations of reality. In order to solve these problems, they negotiate with other actors in order to obtain an agreement on the way technology should develop.

Thus the social and historical studies of technology have hinted at, but not fully explored, the covariation between technology and intra- and inter-organizational development. Hughes draws the conclusions of his study of the development of the electrical power industry in terms of the evolution of the industry as a seamless web of technical and social features (Hughes, 1983).

Our research can be located at the junction of these three approaches. Vergragt (1988) has put forward an analytical scheme to account for the dynamics of internal coalition-forming in relation to the dynamics of technological development. He argues that in several stages of the innovation project, crucial decisions have to be taken concerning continuation, change of direction, or investment of money. These decisions are made necessary by crucial problems arising either from the external environment (changes of market relationships, government activities, outcome of patent lawsuits, etc.), or brought about internally (failure of the technology to perform, organizational difficultures, etc.).

Mulder argues that the coalition-building around the creation of a new technology can be described and analysed in network terms (Mulder *et al.*, 1989, 1990). Analytically, coalitions can be described as network linkages between actors, both inside and external to the firm. The network linkages may be resource linkages, coordination linkages, or communication linkages; it has been claimed that in this particular case of technology development, the network linkage would consist of a mutual agreement on the action perspective (Mulder *et al.*, ibid.). But at the level of historical detail, the dynamics of the internal organizational processes are metaphorically linked to both the external organizational and technological position of the firm. In the case studies that will be elaborated below, an important point appears to be that technical and scientific personnel were able to exploit the meaning of "strong" in strong fibre as a double-eged sword. Not only was the fibre itself physically strong, it also appealed to the preoccupation of higher management at that time to find a "strong" product to help the division survive. Thus the symbolic nature of this particular product is important, an argument that would fit with the social-constructivist and historical approaches.

Analysis of the network structures and their dynamics should be able to account for the dynamics and changes of the technical development. This, however, goes beyond the scope of this chapter.

In sociology-inspired explanations of technological development, such as those above, economic arguments remain rather obscured by the interests, goals, and motives of the actors. Clearly, the motive for technology development and technological innovation is an economic one: as Schumpeter pointed out, big industrial companies compete in creating new technologies. Similarly, we can argue that the behaviour of managers, to act as a product champion, or to form internal and external coalitions, is clearly stimulated by economic needs. The perceived economic needs and opportunities of the firm are the most important variables driving the activities of managers and scientists, and thus the creation of new technologies.

However, the nature of this relationship between technology and economy is not at all clear. Economic theory does not help us much further on this point. Perhaps we should start by saying that it is necessary and fruitful to investigate how economic motives are transformed by social actors into other arguments. Neither of these theories encompasses all the aspects of company decision-making on technological innovation. Here we argue that in order to understand firm behaviour with regard to technological innovation, one has to include economic, organizational, and sociological dimensions in a framework for explanation. We will explore the usefulness of this framework by elaborating a few cases concerned with roughly similar technologies but with somewhat different characteristics on each of the three dimensions. Before we introduce the cases, we shall elaborate the three dimensions.

DIMENSIONS OF TECHNOLOGICAL INNOVATION

We consider innovation to be the rearrangement in novel ways of technical, scientific and organizational elements. The degree to which each of these elements plays a role depends on three variables: the type of technology, the type and size of the organization, and the firm's place in its own industry and the characteristics of the market (see also Lawless and Finch (1989), and Mowery and Rosenberg (1989)).

These variables are important if we try to understand different strategies of industrial firms *vis-à-vis* technological opportunities and constraints. In our view, technological innovation by firms is accomplished by processes of choice making. Companies make these choices on the basis of their strategy, which is a combination of an assessment of their own organizational and technological capabilities, and the market opportunities. Thus in retrospect we can analyse a firm's performance in a certain technological field as the result of a choice-making process, in which various elements have played a role. We call these elements the dimensions of the innovative decision space. We distinguish three dimensions which form a framework in which opportunities and constraints can be positioned.

The first dimension is the technology itself. Technology means knowledge about technical artefacts, machines, production processes and natural phenomena, and about the limits these impose on product characteristics. Companies have a certain technological knowledge base which they exploit in the process of technological innovation. They may invest extra resources to obtain new elements of a technology. However, the possibilities are in part limited by the availability of resources for specific projects. Another constraint on the technology is that not everything is technologically possible; some technical constraints cannot be overcome by research.

The second dimension is the organizational structure of the firm. By this we mean not only the formal organization, but also its resources, information flows, and decision-making traditions. More specifically, research and other information-gathering traditions are embedded in this wider structure. The organizational structure may or may not be sufficient (technically or with regard to resources) to support a specific type of technological innovation.

The third dimension consists of the external environment of the firm. By this we mean the competitors, the suppliers, and the consumers, and also legislative regulations, the trade unions, and the environmental movements. Thus, the environment of the firm consists of both market and non-market relations. The external environment plays a role in the economic analysis, because it is relevant for the necessary return on investment for the process of new technology creation. However, economic arguments about continuation, stopping, or modifying a technological innovation are embedded in social processes with an intra- or inter-organizational character. The firm is connected with other firms that are suppliers of equipment and supplies. They can help to overcome specific technological problems, and can provide opportunities for innovation (Håkansson, 1989). Similarly, customers can play a role in the development of technology. The characteristics of competitors and their behaviour are crucial elements of the firm's environment. Non-market interactions can, for instance, consist of R and D cooperation with government institutes and universities, as well as the support or constraints offered by the national environment. Thus neither internal nor external processes are limited by economic factors alone (Granovetter, 1985).

Within the three dimensions mentioned, the art of innovation consists of dealing with the typical opportunities and constraints that the innovative decision-space provides. By acting in specific ways in the various innovation processes, large corporations learn to deal with each of these three dimensions and their various combinations. By using this competence they can create new technologies that grant them some degree of monopoly power. However, apart from very rare occasions, they cannot control these aspects for longer periods of time. Specifically, this last aspect is an essential ingredient of the mechanism for continued innovation.

The creation of an innovation is thus partly determined by the pre-existing internal and external interactions of the firm. However, the development of a new technology itself creates new social structures and reshapes old ones (Tushman and Anderson, 1986; Anderson and Tushman, 1990; Stinchcombe, 1990). We will use the concept of networks rather loosely as an analytic tool to describe how social actors interact during an innovation process, specifically discussing networks of people, organizational units and external organizations that are "connected" to the series of problems that the creation of new technology poses in each of the three dimensions. In order to search for solutions to specific problems, new organizational arrangements may be formed, which thus support the technology. Such arrangements might be called investment groups (Stinchcombe, 1990). We should stress that this model does not set out to be a general scheme to explain all technological change processes. Its intention is to provide an analytical framework for the analysis of organized innovation in large corporations and complex technologies. (Pavitt (1984) has argued extensively that differences between technologies should be taken into account in order to make a satisfactory analysis of technical change.)

This chapter investigates the introduction of new products in the chemical processing industry. Within this industry we selected cases mainly from the synthetic fibre industry. The products are usually for multi-million dollar industrial markets, i.e. intermediate products; they are not consumer or other end-products. Because of their long lifetimes, the replacement of old products by new ones has a far greater impact than innovations in industries where products are characterized by short innovation cycles, or where relatively small markets are dominant.

In the fibre industry, replacement of a product easily leads to a significant change in internal and external market relations. Equally, however, it has been noted that there is a high degree of stability in the industry. In another study of the synthetic fibre industry, it was concluded that early entrants had an advantage over late entrants. The usual explanation given for this effect is that of the learning curve, which can function as a barrier to entry for later producers (Lieberman, 1989; Giacco, 1986). A firm that is able to introduce a major innovation in the petrochemical or fibre industry potentially has in its hands the instrument of either lucrative returns or disaster (Stobaugh, 1985). There is a situation in which conflicting demands along several dimensions have to be met at the same time. In order to obtain the full benefit from a high-cost innovation, knowledge about its details should be guarded from competitors; yet it is sometimes crucial to be able to exploit weaknesses in the technology strategy of other firms. Moreover, technical details have to be discussed with customers who sometimes also buy from the competitor. Of course, when a market changes, users are forced to change their production line and treatment of the product, and other suppliers with a completely different background also have the opportunity to step in.

THE CASES

In this section we discuss five case studies which have been studied in our group at the University of Groningen, some details of which have been published elsewhere and are part of a forthcoming dissertation. The case studies have been selected as differing in one or two dimensions, the other dimension(s) being roughly similar.

First, we introduce two case studies on the development of new polymer-based materials in the Dutch-German AKZO company. The comparison of the TENAX case (Vergragt *et al.*, 1989) with the TWARON case (Mulder *et al.*, 1989, 1990) is illustrative, because the Tenax project case failed, while the Twaron case was a technological success. Both projects were at the time considered to be crucial for the survival of the company.

Second, we compare the AKZO Twaron project with the development of Du Pont's Kevlar. The interesting point here is that these two fibres are technologically similar, but that the companies followed different strategies, eventually leading to a fierce patent struggle between AKZO and Du Pont. Full details of the development of Kevlar appear in a forthcoming PhD thesis by Karel Mulder. A short description can be found in Mulder and Vergragt (1989).

Third, we compare the successful AKZO and Du Pont aramid projects with cases where similar projects failed, i.e. Bayer and Monsanto. From these comparisons we should learn more about the relative importance of each of the three dimensions of the success or failure of innovation projects.

AKZO's development of Tenax: the project that failed

Technology
In 1960, General Electric published a patent for the production of polyphenylenes. AKZO was interested in new polymeric substances with which to produce fibres and electrical insulation foils, and started research, which led to the granting of an application patent.

The first steps in the further development involved establishing a research-oriented joint venture with GE. The applications of the material that were sought were in the high performance fibre area and as a synthetic high voltage insulation paper for high voltage transmission cables. Two specific materials were under consideration: first PPO and later on P30. At the end of the 1960s it became clear that PPO was inappropriate for both applications, and the joint AKZO-GE effort was terminated. (GE Plastics later developed a PPO/polystyrene blend, which it marketed as Noryl.) AKZO went on to study the use of P30 as a possible material for the insulation of extra high voltage power lines. This project was called Tenax. Expectations were high, because of the expected growth of energy use and transmission at the time. The research concentrated on the preparation of P30 synthetic paper with low conductivity and high strength. Problems occurred in achieving mechanical as well as electrical properties, however; the values that were predicted by a theoretical model could not be attained. This led to the eventual collapse of the project in 1973.

Internal organization
The Tenax project was internally organized in a way which was rather new for the time (1970). A kind of venture firm was formed, in which the R and D and marketing functions were brought together. This group operated fairly autonomously from the rest of the firm, and was supervised directly by an AKZO board member. Although morale in this group was high in the beginning, it seems that the eventual failure was also a consequence of this organizational innovation. Because of the organizational structure, which was new to AKZO, it seems that both the integration in, and the control by the large company were insufficient.

Market and competition
The Tenax project was a joint effort by AKZO, KEMA (the joint research institute of the Dutch power companies, responsible for the application research for the insulation of cables), and Dainichi, a Japanese producer of cables. Dainichi developed processes to wrap the paper made by AKZO around the cables. However, as an insulating material, Tenax had to compete with non-synthetic paper. Challenged by the possible development of synthetic paper, and by rising demand, the craft paper producers were able to improve the electrical properties of their material. Moreover, there appeared clear signs of diminishing economic growth and energy shortages, which became very clear during the 1973 oil crisis.

Thus, the AKZO, KEMA and Dainichi consortium failed to deliver the technical properties, and were faced with decreasing market expectations and increasing technical competition by the craft insulation-paper makers. The combination of these factors proved fatal for the project.

AKZO's development of Twaron: the successful imitation of a competitor

Technology

The AKZO fibre division operated in industrial fibre markets, making tyre cords and yarn for tyre reinforcement. The announcement by Du Pont in 1970 that it had developed a fibre with far better mechanical properties, Fibre B, which was later called Kevlar, was a threat to AKZO's market position. The scientists of AKZO knew that Du Pont was working on aramid fibres (see Black and Preston, 1973). In 1973, they were able to produce their own aramid fibre. The technology appeared to parallel closely that developed by Du Pont; therefore the patent position of Du Pont was the main obstacle during the mid-1970s. To get around Du Pont patents, AKZO scientists tried to develop their own process to make the polymer. They succeeded in finding a new solvent by which the polymer could be made, which they patented in 1975.

Internal organization

After the Tenax failure, AKZO did not carry out any experiments which would prevent the integration of R and D projects. A close analysis of the project shows that internal coalitions were formed between the research laboratory, the patent bureau, the engineering department, and some proponents on the executive board. This coalition, or network, supported the project even at times when its future was at stake. One of these moments was when the whole company was in danger of collapsing in the face of a tremendous market crisis in 1975, and Du Pont refused to license AKZO for its Kevlar process. The internal coalition was able to develop enough power to continue and eventually to convince the board of directors to challenge Du Pont's patent position.

Market and competition

The internal coalition or network was supplemented by an external network: The Dutch Department of Economic Affairs, together with the Northern Development Company, were successfully approached to support further commercialization of the aramid fibre. This external coalition was necessary to provide the funds for the investment in production units. At the same time, a patent conflict with Du Pont broke out, and AKZO needed government support for the struggle to enter the American market.

The external network was strong enough to provide the funds for investment, and the patent conflict was solved in 1988 with an agreement between AKZO and Du Pont. However, the market for aramid fibre has grown more slowly than expected. It remains to be seen whether the production will become a commercial success. The break-even point was reached at the end of 1990.

Du Pont's development of Kevlar

Technology

As part of a large-scale programme to develop new high-performance fibres, in 1965 a Du Pont scientist discovered a way to produce a high-strength and inelastic aramid fibre. The fibre was further developed, and introduced at the beginning of the 1970s as "Kevlar". Competitors had been working on similar fibres during the 1960s, e.g. Monsanto in the US, and European companies such as Bayer and Rhône Poulenc, and some time later ICI and AKZO.

The first application that was envisaged was as a tyre cord. Tyre quality had to be improved, to keep pace with increases in car size and speed. The technical development of Kevlar fibre was successful, but at the same time Michelin introduced steel-wired radial tubes: *the* commercial success in the tyre industry. In 1975, further commercialization of Kevlar was postponed, and other applications had to be developed. At the same time, it appeared that the solvent used in the production process was carcinogenic.

Patents on the product itself and the spinning procedures were considered to be the main competitive advantage held by Du Pont. However, a different solvent had to be found. In this area AKZO had an advantage, because it had discovered another solvent which it had patented. Eventually this was one of the main reasons why Du Pont had to reach an agreement with AKZO.

Internal organization

In the 1960s, Du Pont relied heavily on its R and D capability. Its success with nylon was the shining example for Du Pont's scientists, encouraging corporate management to provide staunch support for technological innovation. The development of a strong and inelastic fibre had been a long-term aim of the Du Pont Fibres Department since 1948. The pioneering laboratory of this department could therefore work on long-term projects to reach this goal. Pioneering scientific research in a completely new field ensured Du Pont of a few very important basic patents. Subsequently, application research was set up together with tyre producers and other potential customers.

Market and competition

The same remarks apply as have been made in the AKZO case. Du Pont is the largest producer of aramid fibre in the world and, until the mid-1980s, enjoyed a monopoly position. In the near future it will have to face other competitors besides AKZO in the market, particularly when patents have expired; but in a growing market this will probably not be a serious problem. Moreover early producers benefit from experience gained in established plants. The financial break-even point for Kevlar was reached in 1986.

Bayer's failure with aramid fibre

Technology

In the 1950s and 1960s, Bayer was a producer of bulk fibres for textiles. In 1964 Bayer entered aramid technology by starting a project to support its growing interests in the fibre field. In 1967 Bayer chemists at the new laboratory at Dormagen synthesized an aramid fibre with very good heat-resistant properties. However, the fibre had to compete with polyester. The polyester price was going down rapidly at the time, and it gradually appeared that polyester would be impossible to match in this respect.

By coincidence, in 1969 a process was found for making a strong fibre. Bayer obtained a patent in 1971, but after Du Pont's introduction of Kevlar, Bayer stopped the project. The corporation did not manufacture tyre cords. It had only intended to do so if the fibre that was found had been unique. However, no one would want to compete with Du Pont in an unknown market, with an untried technology.

Internal organization

The most significant event in relation to Bayer's activities in the aramid fibre field was the creation of a new fibre laboratory at the company's Dormagen site in 1964. In this spacious new laboratory there was room for new projects. Thus the relatively eccentric siting of the laboratory here was an important factor for the possibility of starting this project. The peripheral geographical position, however, was also one of the reasons for the eventual failure: the distance between the laboratory and the corporate board meant that the project did not obtain enough support from the corporate management.

Market and competition
The main reason for the failure of the Bayer strong aramid fibre was that Bayer was in the textile fibre market, not the industrial fibre market (where the product needed to be highly temperature resistant and/or very strong). Thus, the researchers did not receive the support of the marketing department. The fact that Du Pont entered the market was an important additional factor for Bayer not to continue the aramid fibre project.

Monsanto's failure with aramid fibre

Technology
In 1966 Monsanto scientists found a way to produce a very strong fibre. However, the Monsanto Textile Fibres Division was not interested, for it could not be used as a tyre cord, the main market for strong fibres. When a market crisis struck the Textile Fibres Division in 1967, some textile fibres research groups were turned over to Monsanto's New Enterprise Division. This division was interested in developing a fibre for reinforcement of engineering composites. Pilot plant production of the fibre, called X-500, started.

However, within the New Enterprise Division, the X-500 project had to compete with the development of carbon fibres, which were also used for reinforcement of composites. An internal struggle started, which was lost by X-500. Monsanto marketing staff did not want to develop a new market for X-500, because "there was nothing like it on the market". In 1968, the effort was terminated. "If we had known about Du Pont's Kevlar, we would perhaps have been able to keep our project going."

Internal organization
This R and D project took place in an isolated laboratory in Durham, North Carolina. There was almost no integration into the company. Moreover, the laboratory, as well as the fibres division of which it was a part, had been fighting to maintain its independence from the corporate headquarters in Saint Louis. The fibre division had a long tradition of struggle with the corporate headquarters (see Forrestal (1977)). Internal division therefore prevented the establishment of any group at a high level in support of this particular technology.

Market and competition

Monsanto was the fourth-largest chemical company in the USA, and the second-largest fibre producer. The main reason for Monsanto not continuing with the aramid fibre project was internal opposition to the project, stimulated by a tradition of being second on the market.

CONCLUSIONS FROM THE CASE STUDIES

We will address two points, before going into a detailed discussion of each of the dimensions we introduced. The first is the (perceived) success or failure of the technology. The second is the question as to whether the technology was developed according to the original goals, or underwent a significant transformation.

As to the success or failure, introduction into the market can be used as a measure. Only AKZO and Du Pont have been more or less successful in introducing a new strong fibre on the market. Even for these companies, the success has certainly not been as great as that anticipated. The AKZO Tenax project was a complete failure, as were the Bayer and Monsanto aramid fibre projects. What is interesting in these cases is that the successes and failures were brought about by different mechanisms, which we will discuss later.

As to the second aspect, only in the case of Du Pont was the technology more or less developed according to plan. However, a detailed analysis of the market, preceding the investment decision, proved to be wrong. Instead of tyre belts, Kevlar is mainly used for military equipment such as helmets and bullet-proof vests. Composites with kevlar fibres are also used in aerospace applications. Because both markets are high-end and protected, in which direct competition is less effective, the success can be deemed only partial. In the case of AKZO's aramid fibre, the same problems played a role, but the production technology had to be changed for two other reasons. The first was the strong patent position of Du Pont, which forced AKZO to invent new steps in its production technology. The second was the disovery of the carcinogeneity of a solvent, which suddenly forced Du Pont to reach a late agreement with AKZO. Neither of these factors were technology-inherent.

The technology

The development of technology can be characterized as offensive, defensive, or marginal. In the case of Du Pont, there was a case of offensive development of a new technology. Du Pont had the scientific and technical expertise to develop a completely new technology, based only on the accumulation of technical expertise, the firm's success with nylon in the past, and an internal organizational structure that supported its development.

In the case of AKZO's Tenax, the failure in the technology was the result of an accumulation of factors. The knowledge base to develop the new technology was insufficiently developed, the market was new and unknown, new joint ventures for research and marketing had to be set up, and there was a new organizational experiment.

In the case of AKZO's Twaron, there was a defensive, or following, strategy. AKZO was unable to invent the new technology itself, but was able enough to circumvent the patent barriers posed by Du Pont. Also, they knew the market very well, and they reversed their organizational innovations.

In the case of Bayer and Monsanto, there were clearly signs of marginality with respect to core business. In both cases, strong fibres were not considered to be the core business. In Bayer's case, the organization of the project was such that it was geographically and organizationally marginal. Moreover, strong fibres were not considered to be part of Bayer's business, because Bayer was in textile fibres. In the case of Monsanto, the project was performed by a part of the firm that did not wish to become integrated in the mother firm, and thus remained peripheral.

The internal organization

The internal organizations of the companies studied can also be classified according to different characteristics. In the case of AKZO's Tenax, an experimental venture organization was coupled loosely to the main organizational structure of the firm. Although in some cases, for instance in a bureaucratized organization, such a structure may prove to be beneficial, in this case it clearly failed. This is apparent for two reasons. In the first place, the project collapsed in a very brief time span, in contrast to AKZO's Twaron which survived a number of major crises of the company. The other reason for the survival of the AKZO Twaron project was because it proved to be rooted very deeply in the AKZO organization. The AKZO Twaron project thus was based on an internal coalition of powerful actors which remained intact even in the case where the management once formally decided to stop the project.

Thus AKZO's Twaron can be characterized as a strong internal coalition. In contrast, the internal structure at Du Pont can be characterized as strong leadership in combination with a powerful corporate culture which supported the idea of offensive research. The internal structure at Bayer can be characterized as a very loose coupling (both geographically and organizationally) to the central management. This made the project possible in the first place, but was also one of the causes of its eventual failure. The internal organization of Monsanto similarly was characterized as very loosely coupled to the central management.

To conclude, it appears that the form of internal organization is at least one of the main determinants of the success or failure of a large technological innovation. Although this conclusion may sound rather superficial, it should not be forgotten that this aspect is often overlooked in all except the managerial theories of large-scale technological innovations. Although the internal organization is crucial, there is by no means a unique solution to the organizational question, as has been shown by the differences between the AKZO and Du Pont cases.

The external factors

The significance of different external factors differs from case to case. Clearly, in the case of a very large firm like Du Pont, external factors affect various operations in a different manner from that in a somewhat smaller firm such as AKZO. However, in Du Pont's case the success of the innovation (or the non-failure) depended on market forces. Thus the military market saved the Kevlar project, which would probably have failed had it had to rely on the civil market alone. The other important external factor was the patent struggle with AKZO. Du Pont was clearly protected in this struggle, both legally and politically.

For AKZO the situation is somewhat more complicated. In the Tenax case, AKZO was very dependent on collaboration with other firms like KEMA and Dainichi. The lack of strength in this coalition was another cause for failure. In the Twaron case, AKZO was very careful to build up relationships with customers, with government agencies (both for funding and for the necessary licences), and with central government (for support in the patent struggle). These coalitions proved to be strong enough to survive over a prolonged span of time.

In the cases of Bayer and Monsanto it can be shown that external coalitions did not play an important role, because these firms did not reach the stage of commercialization. It can be argued that the commercial market was not matched with the technology: Bayer was in textile fibres, and Monsanto in high temperature resistant fibres.

To conclude, we see that external relations (both market and non-market) play an important role, but mainly in the later stages of the innovation.

GENERAL CONCLUSIONS

Internal coalitions and technology development

The existence of support within the firm is not so much a result of careful planning, as of building strong relations of trust. In part, internal coalitions are built on the real or perceived resources provided through assessment of the external environment of the organization (see also Daft and Weick (1984) with regard to external perceptions and their influence on corporate strategy). The essential part is that during technical change brought about by large-scale innovation, none of the existing "factual" relations can be trusted. Thus input by the marketing department does not concern real markets, but rather perceived markets and their needs. The technological position of the firm is dependent not only on its internal technical capacity, but also on the legal and technical position of its competitors.

What is the reason such external factors — perceived or real — influence internal coalition-building in large-scale technology development to such a large extent? Apparently the main reason is that the organization has to commit resources to the next stage, and testing the possibility of success is essential to internal decision making. At the first stage of a core technical project, externalities influence the course and duration of a project.

In the introduction we mentioned Vergragt's conceptual schema. As a consequence of these case studies we add to this schema the following consideration: that in most cases these crucial decisions are taken by the coalition or network that supports the technology. This is clearly illustrated in the Tenax case, where the termination of the project was brought about by the sudden collapse of the supporting coalition.

To summarize this section, the development of a new technology by a firm is accompanied by the formation of internal and external coalitions of actors; and the analysis of the social structure of these coalitions provides insights to the dynamics and direction of technology development.

Inter-organizational relations and technological development

As can be concluded from the comparison between projects in aramid fibres that succeeded and those that failed, commitment of resources dovetails with management commitment. This is understandable, because large-scale innovation in this industry shapes organizational strategy for a long period of time. Therefore decisions on technology in such cases not only reflect the technical aspects of the innovation, but also other reasons to continue. In the decision, the internal failure precedes an external failure.

When Bayer and Monsanto are compared to Du Pont and AKZO, commitment seems to be one of the main factors involved. Du Pont and AKZO were well organized in the field of industrial fibres. Du Pont, in that period still adhered to a tradition of radical innovation of large-scale products (Hounshell and Smith, 1989). AKZO had a need for high-tech fibres that would be capable of helping the fibre division to protect its market share, notably in the tyre market segment. Both corporations transformed the projects into central projects for the survival of the firm strategy. Thereby the necessary internal support was created to succeed with these mega-projects.

Commitment in Bayer and Monsanto emerged in relation to a business sector that did not have a central position in the main strategy of either firm. In themselves, these differences seem to explain the success of the first two and the failure of the second two cases. But when we follow the unfolding of the conflict of Du Pont and AKZO, and the actual use of the strong fibres that were produced, a different story emerges about the predictability of success. Monsanto and Bayer had potentially much larger resources to support the fibre development. They were early entrants in the field. Neither had the internal supportive capacity to carry on with those projects. With hindsight, the fibre activity was a peripheral concern, either as a consequence of internal cultural clashes (as in the case of Monsanto) or as a consequence of an unstable laboratory position (as in the case of Bayer).

Generally speaking, the possibility of establishing relations with customers is an essential prerequisite to the introduction of a technology. Established, well-defined ties with customers are important in order to retain the market position (cf. Stinchcombe (1990), Leonard-Barton (1988)).

Characteristics of technological development

Technological development thus is made possible through the creation of internal coalitions or networks, and by extension of these networks to include other organizations in the environment. These relations might in the end become economic market relations. For instance, in the aramid case it was clearly an analysis of the market and the firm's position in it that brought AKZO to the development of aramid fibres. However, the initial market estimates turned out to be far from reality. In the AKZO case, there have been several opportunities where, from a clear economic perspective, the decision should have been taken not to continue the project. Clearly, other motives and mechanisms played a role.

We do not argue here that these mechanisms are purely sociological and not economic in nature. It may be an economically sound principle to motivate the scientific workforce by giving them a largely free hand. In this respect, the sudden stopping of an innovative project for economic reasons that are shrouded to a large extent in uncertainty, could frustrate the participants and therefore impede future technological capabilities of the firm.

Thus we imply that most actors inside and around the firm act on economic grounds, but translate these into other arguments and interests. It seems to us a task for studies of technological innovation to pay more attention to these translation processes.

On the specific case of large-scale technologies, we can conclude that the model of mutual shaping of the technology and the social structure is appropriate. Two phases of technology development can be discerned. In the first phase, contingencies external to the organization are interpreted by various actors within the organization. The perception of strength and weaknesses of the organization is much more important for the strategy choice than the fact-based evaluation of the external world. In the second phase of the innovation the external relations of the organization have to be developed in a manner that suits the specific character of the technology. Non-market forces, like defence or national technology policies, are a significant part of these alliances. Thus, generally the firm has to select between options to ensure continued survival. This selection process involves changing the technological environment. Environmental change in its turn is not manageable in a similar manner to internal change in the organization (see also Anderson and Tushman (1990), Stinchcombe (1990)).

The three-dimensional approach applied to a set of comparable case studies enabled us to show that our conceptual scheme accounts at least for the salient features of technological innovation in each of these five examples. To conclude, we would like to state that the current segmented theories of technological innovation may all be plausible because they attend only to one aspect of the innovation process. In reality, the internal decision making in firms hinges on the assessment of the organizational, economic and technological factors themselves, but crucial aspects are hidden in the interrelation between the three dimensions. We have shown by our case studies that the ensemble of these factors, which we called the innovative decision space, accounted for success and failure, as well as major changes in large-scale technological or service characteristics of specifically chemistry-based technology. We think that future work should extend the notions sketched in our conceptual scheme.

References

Anderson, P. and Tushman, M.L. (1990), Technological Discontinuities and Dominant Designs: A Cyclical Model of Technological Change, *Administrative Science Quarterly*, **35**, 604–33.

Bijker, W., Hughes, T., and Pinch, T.J. (Eds) (1987), *The Social Construction of Technological Systems*, MIT Press, Cambridge, MA.

Black, W.B. and Preston, J. (eds) (1973), *High modulus wholly aromatic fibers*, New York.

Callon, M. (1986), The Sociology of an Actor-Network: the Case of the Electric Vehicle, In *Mapping the Dynamics of Science and Technology* (Eds M. Callon, J. Law and A. Rip), Macmillan, London.

Coombs, R., Saviotti, P. and Walsh, V. (1987), *Economics and Technological Change*, Macmillan, London.

Daft, R.L. and Weick, K.E. (1984), Toward a Model of Organizations as Interpretation Systems, *Academy of Management Reivew*, **9**, 284–95.

Forrestal, D.J. (1977), *Faith, Hope and $5,000, the story of Monsanto*, New York.

Giacco, A.F. (1986), Meeting Challenge in the Chemical Industry, *Research Management*, **29** (1), 36–9.

Granovetter, M.A. (1985), Economic Action and Social Structure: The Problem of Embeddedness, *American Journal of Sociology*, **91**, 481–510.

Håkansson, H. (1989), *Corporate Technological Behaviour, Co-operation and Networks*, Routledge, London.

Hounshell, D.A., and Smith, Jr, J.K. (1989), *Science and Corporate Strategy, Du Pont R and D, 1902–1980*, Cambridge University Press, Cambridge.

Hughes, T.P. (1983), *Networks of Power, Electrification in Western Society, 1880–1930*, Johns Hopkins University Press, Baltimore, MD.

Law, J. and Callon, M. (1988), Engineering and Sociology in a Military Aircraft Project: A Network Analysis of Technological Change, *Social Problems*, **35**, 284–97.

Lawless, M.L., and Finch, L.K. (1989), Choice and Determinism: A Test of Hrebiniak and Joyce's Framework on Strategy-Environment Fit, *Strategic Management Journal*, **10**, 351–65.

Leonard-Barton, D. (1988), Implementation as Mutual Adaptation of Technology and Organization, *Research Policy*, **17**, 251–67.

Lieberman, M.B. (1989), The Learning Curve, Technological Barriers to Entry, and Competitive Survival in the Chemical Processing Industries, *Strategic Management Journal*, **10**, 431–47.

Mowery, D.C., and Rosenberg, N. (1989), *Technology and the Pursuit of Economic Growth*, Cambridge University Press, Cambridge.

Mulder, K.F., and Vergragt, P.J. (1989), "Weaving the Superfibre Network, or how AKZO developed its aramid fibre", paper, Groningen, August.

Mulder, K.F., and Vergragt, P.J. (1990), Synthetic Fibre Technology and Company Strategy, *R and D Management*, **20** (3), 247–56.

Nelson, R.R. and Winter, S.G. (1982), *An Evolutionary Theory of Economic Change*, Belknap, Cambridge, MA.

Pavitt, K. (1984), Sectoral Patterns of Technical Change: Towards a Taxonomy and a Theory, *Research Policy*, **13**, 343–73.

Pinch, T. and Bijker, W. (1984), The Social Construction of Scientific Artefacts, *Social Studies of Science*, **14**, 399–441.

Stinchcombe, A.L. (1990), *Information and Organizations*, University of California Press, Berkeley.

Stobaugh, R. (1985), Creating a Monopoly: Product Innovation in Petro-chemicals, *Research on Technological Innovation, Management and Policy*, **2**, 81–112.

Thorelli, H.B. (1986), Networks: Between Markets and Hierarchies, *Strategic Management Journal*, **7**, 37–51.

Tushman, L. and Anderson, P. (1986), Technological Discontinuities and Organizational Environments, *Administrative Science Quarterly*, **31**, 439–65.

Twiss, B. (1980), *The Management of Technological Innovation*, Longman, London.

Vergragt, P.J. (1988), Social Shaping of Industrial Innovations, *Social Studies of Science*, **18**, 483–513.

Vergragt, P.J., Mulder, K., Rip, A. and van Lente, H. (1989), *De Matrijs van Verwachtingen Ingevuld voor de Polymeren Tenax en Twaron*, Verslag van een Studie in Opdracht van de Nederlandse Organisatie van Technologisch Aspectenonderzoek, Twente, Groningen.

White, H.C. (1981), Where do Markets Come From?, *American Journal of Sociology*, **87**, 517–47.

Williamson, O.E. (1986), *Economic Organization*, Harvester Wheatsheaf, New York.

INDEX

DATE DUE   L.-Brault